This book reviews the long-standing debate over the relative merits of a high-protein versus a low-protein diet. When protein (or "animal substance") was first discovered in vegetable foods, it was hailed as the only true nutritional principle. Justus Liebig, the leading German chemist of the mid-nineteenth century, believed that it provided the sole source of energy for muscular contraction. In contrast, health reformers argued that high intakes were overstimulating, leading to dissipation and decline. U.S. government publications in the 1890s recommended that working men receive 125 grams of protein per day, but work at Yale indicated that men maintained their strength on half that intake.

In the 1950s kwashiorkor, a disease of infants in many Third World countries, was judged to be the result of simple protein deficiency. The United Nations declared a world protein shortage. But the causes of kwashiorkor were reassessed, and projects to produce novel protein sources were eventually abandoned. Today there is again concern that overconsumption of protein in affluent societies may damage health. This book puts the protein controversy into a historical perspective that sheds light not only on the subject itself, but on the scientific process as well.

Protein and Energy

Protein and Energy

A Study of Changing Ideas in Nutrition

KENNETH J. CARPENTER
University of California, Berkeley

CAMBRIDGE
UNIVERSITY PRESS

University Printing House, Cambridge CB2 8BS, United Kingdom

One Liberty Plaza, 20th Floor, New York, NY 10006, USA

477 Williamstown Road, Port Melbourne, VIC 3207, Australia

314-321, 3rd Floor, Plot 3, Splendor Forum, Jasola District Centre, New Delhi - 110025, India

79 Anson Road, #06-04/06, Singapore 079906

Cambridge University Press is part of the University of Cambridge.

It furthers the University's mission by disseminating knowledge in the pursuit of education, learning and research at the highest international levels of excellence.

www.cambridge.org
Information on this title: www.cambridge.org/9780521452090

First published 1994

A catalogue record for this publication is available from the British Library

Library of Congress Cataloging in Publication data
Carpenter, Kenneth J. (Kenneth John), 1923–
Protein and energy : a study of changing ideas in nutrition /
Kenneth J. Carpenter
p. cm.
Includes bibliographical references and index.
ISBN 0-521-45209-0 (hc)
1. Proteins in human nutrition. 2. Proteins in human nutrition – Government policy – History. 3. Energy metabolism. I. Title.
QP551.C37 1994
613.2′8 – dc20 93-32130
CIP

ISBN 978-0-521-45209-0 Hardback

Contents

Tables and Figures

Tables

Tables and Figures

Figures

Preface

THIS BOOK HAS ITS ROOTS in my desire to understand, first, how the idea arose, and became endorsed by the United Nations Organization in 1965, that a growing nutritional problem requiring immediate attention was a worldwide shortage of protein and, second, what caused the idea to be suddenly discarded, amid some passionate debate and name calling. Although it is of considerable importance that such an about-face be understood, and appropriate lessons drawn from it, the nutrition community seems largely to have ignored the subject as an embarrassment.

However, this controversy from the 1950s and 1960s was really the third of a series of disputes on whether human diets commonly contained either too little or too much protein – the first coming in the mid-nineteenth century and the second at the beginning of this century. I have therefore tried to set out the development of ideas about protein from their recognizable beginnings so that the whole sequence will appear in a single volume.

Of course, in the seventeenth and eighteenth centuries the terms "protein" and "energy" did not exist, but investigators were struggling toward the idea that while one part of our food was used to replace worn-out tissues, or provide the material for growth, another part provided "fuel for the flame of life" and was the source of animal heat. At first, after protein was discovered, it was thought also to be the source of mechanical energy. Although this notion was disproved, investigators still thought that a high intake of dietary protein was responsible for nervous energy in the sense of vigor, or "get up and go." One school of thought held that the relative prosperity of people in Western countries was a result of their vigor, which was dependent in turn on their eating substantial quantities of meat. Others believed that this practice was overexciting and stimulated uncontrollable sexual energy, which had debilitating and demoralizing effects, so that a low-protein, vegetarian diet was to be preferred.

Particularly in the earlier years, the key advances came from only a small number of people, and I have tried to say something about their interests and backgrounds at times when there was no established career pattern for someone wanting to work in this field. New ideas in science do not just happen: they are products of individual minds, influenced by specific backgrounds. And it is this personal side of the history of science that increases its interest, at least for some of us.

The historical chapters have all been written, to the best of my ability, in terms of the problems and questions *as seen by investigators at the time,* and without an immediate reinterpretation of their observations from a modern point of view. Such an approach seems the only way for us to try to put ourselves in the shoes of our predecessors. Readers with an interest only in modern nutritional concepts could begin at Chapter 6, but I hope that they will, in time, work their way back to the origins of the subject and come to respect and appreciate the contributions made in the earlier periods.

Although this book is intended to be a serious study of the development of ideas about the human diet against the background of general scientific concepts in each period, I have tried to put the issues in terms that will not deter the general reader. At some points it has seemed necessary to explain in detail, for the sake of those interested, how authors did their calculations to arrive at their estimates of protein needs, but the reader can skip these without losing the thread of the arguments. Because "nutrition" is something that involves us all in everyday decisions, and because its scientific study has not, in general, been very mathematical or dependent on abstruse concepts, it is an area in which nonspecialists can study the gradual development of scientific methods of investigation and their application in practice.

Because of the many relevant papers published since 1850, I can only offer a survey of this period. I apologize for omitting the work of some significant contributors. It could be said that a study should be limited to the intensive and more comprehensive scrutiny of a single period of the past. Rightly or wrongly, I have had the different aim of trying to connect the past with the present, and still to use direct quotations from leading individual scientists within each period.

There are many other aspects of dietary protein and related issues that this book does not cover. For example, it is not concerned with the social causes of poverty-related malnutrition or with the design of nutrition policies and their implementation in the Third World. Nor is it concerned with debates on the long-term effects of malnutrition in children or, in another direction, with the biochemical mechanisms that control protein synthesis in the body. All of these are matters of great interest and importance, but others have written about them more knowledgeably than I could. One chapter does discuss scientists' attempts to produce new protein foods as a direct response to the conclusion that there would (or could) be a world protein crisis. Otherwise, the book is concerned solely with the evolution of scientific ideas about the need for protein and how our actual requirements could be determined.

As far as I know, this is the first book to review the development of these ideas over such a long time span. However, any author in such a field is indebted to those who "went before" as specialists in particular aspects of the subject. I feel particularly indebted to Professor J. R. Partington, among the dead, for his four-volume *History of Chemistry* and to Professor F. L. Holmes, among the living, particularly for *Claude Bernard and Animal Chemistry* and his introduction to Liebig's *Animal Chemistry*, as well as for his encouragement.

I am also grateful to my colleagues Marc Schelstraete and George Wolf for assistance with some convoluted German writing from the preceding century, to Doris Calloway and Patricia Swan for their comments on portions of the manuscript and to Frank Smith of Cambridge University Press and his anonymous reviewers, as well as to the production editor, Mary Racine, for their suggestions. Finally, the secretaries in the Department of Nutritional Sciences have put a handwritten manuscript, and many revisions, onto a word processor with skill and good humor, and this has been much appreciated.

1 Nutritional Science before the Chemical Revolution, 1614–1773

THE OBJECT OF THIS BOOK is to trace the origins of and changing ideas about the role of protein in our diet and the quantities needed for optimal health. The word "protein" was not coined until 1838, and the chemistry of materials falling into this class was only beginning to be understood in the preceding 40 years. However, scholars were interested much earlier than this in the basic question of whether animals (and humans) had the power to turn any kind of digestible food into the material of their own tissues, or only certain fractions of it, having basically similar properties. There were also differences of opinion as to whether growth in the young, and tissue replacement in adults, was the *sole*, or even the *major*, function of nutrition.

In classical Greek medicine in particular, "diet" was an important consideration in the maintenance of health, although at that time the word meant the whole manner of one's life and environment, as well as one's food.[1] In the second century A.D., Galen had written: "Our bodies are dissipated by the transpiration that takes place through the pores in our skin that are invisible to us; therefore we need food, in a quantity proportionate to the quantity transpired."[2] Foods were classified primarily according to what were thought to be their immediate effects on a person's mood. One food would be sedative, another aphrodisiac, and so on.[3] Different fōods were recommended for people with different temperaments (hot, cold, dry, moist) in order to bring them more into the norm or "ideal" balance.[4]

Recommendations originally made in the classical period were still being proposed in the sixteenth century. Persons of the choleric type were advised

1. Trémolieres (1975); Fidanza (1979).
2. Cited by Temkin (1960), p. 86.
3. Cosman (1983); Mauron (1986).
4. Eliot (1541).

to eat coarse meat (such as beef rather than chicken), which was only slowly digested, and to have frequent meals; otherwise, the intense heat of their digestive system would scorch and damage the empty stomach, with the fumes rising to cause a headache. By contrast, naturally melancholic persons should eat moist and easily digested, boiled meats and should drink milk.[5] These ideas seem strange and incoherent to the modern reader, but they were in fact built up from a complex framework of assumptions.[6]

The New Beginning

There is little connection between traditional belief and the science of nutrition that was stirring in the seventeenth century on the basis of new developments in physical science and physiology. The first influential publication in this period came from the Italian scientist Santorio (usually known by his latinized name, Sanctorius). He was a professor in Padua, the seat of a famous university but best known to English-speaking people from Shakespeare's references to it in the *Merchant of Venice* and the *Taming of the Shrew*, which were written about the time that Santorio was studying there. Santorio repeatedly weighed his food and drink, on the one hand, and his excreta (urine and feces), on the other, and also measured changes in his own weight. He reported in 1614 that, on average, his daily intake amounted to about 8 lb. and his excreta only to about 3 lb. Since there was no significant change in his weight, there was an unexplained daily disappearance of some 5 lb. of material. At that time respiration was thought to serve only as a means of cooling the heart. And since it seemed that this amount of food and drink was necessary, Santorio concluded, for want of any other explanation, that the daily "disappearance" must be due to the breakdown of this amount of body tissue, which was then excreted largely through the skin in the form of "insensible perspiration." The losses were made good by the nourishment ingested.[7] This was only a more quantitative restatement of Galen's view in the 2nd century A.D., but Santorio's work made a great impression.

The second significant advance was published in 1628 by William Harvey, an Englishman who had obtained his medical degree at Padua. This was the discovery of the circulation of the blood from the left side of the heart throughout the body, via arteries, capillaries, and veins and back to the right side of the heart, with a second circulation to the lungs. Harvey's demonstration of a constant, rapid perfusion of the body's tissues by arterial

5. Boorde (1567).
6. O'Hara May (1977).
7. Santorio (1614); Bylebyl (1977), pp. 377–8.

blood provided a mechanism to explain the extensive movements of old material away from the tissues and the incorporation of new.[8]

The French philosopher René Descartes, in his *Discourse on Method,* first published in 1637 and still regarded as one of the seminal works in the development of scientific thinking, made use of Harvey's finding. He believed that the living heart was extremely hot and argued that it was the repeated exposure (perhaps 100 to 200 times per day) of absorbed nutrients circulating in the blood to this "hot spot" that gradually "distilled" them into the "stuff" of animal tissues. The process might take longer with some foods than others, but all would eventually be converted.[9] This was not a totally original concept, as has been shown elsewhere.[10]

We see a somewhat different view in *The Natural History of Nutrition, Life and Voluntary Motion,* published in 1659 and considered the first book on human physiology to have been written in English (Figure 1.1). The author, Walter Charleton, born in 1620, was exposed to the new ideas of Harvey, Descartes and others as a student at Oxford, where he earned an M.D. degree. He was a fluent writer, in particular trying to explain recent scientific discoveries and to reconcile them with Christianity, but his career had its ups and downs. He sided with the royalists and remained in Oxford when the English civil war began in 1642 and, while still in his early twenties, became one of the king's personal physicians. After the restoration of Charles II, he became a founding member of the Royal Society and president of the Society of Physicians.[11]

In his book on nutrition, Charleton wrote that animal heat continually agitates the minute particles of the body and "dissolves, dispenses or consumes" them, so that there must be a continuous assimilation of equivalent particles from our food. That is the first function of nutrition. The second is to provide fuel (or oil) for the vital flame burning within us, which can be extinguished either by suffocation or by want of sustenance.[12] This seems consistent with the modern view that, as adults, we need food partly for "maintenance" due to constant wear and tear, or turnover, of our tissues and partly for combustion as a source of energy and secondarily of heat.

Charleton went on to say that he did not believe that the different parts of the body needed correspondingly different materials for their restitution. He believed that a principle similar to that found in the white of an egg would be suitable for all, and he used the simile of a common rain supporting the growth of trees of different species.[13] (At that time, before the existence

8. Debus (1978), pp. 66–73.
9. Descartes (1649), p. 79.
10. Mendelsohn (1964); Hall (1969).
11. Rolleston (1940); Gelbart (1971); Fleitmann (1986).
12. Charleton (1659), p. 10.
13. Ibid., p. 9.

NATURAL

HISTORY

Of { NUTRITION, LIFE, and VOLUNTARY MOTION.

Containing

All the *NEW DISCOVERIES*

of *ANATOMIST'S*, and most probable

Opinions of *PHYSICIANS*,

Concerning the

OECONOMIE OF HUMAN NATURE;

Methodically delivered in

EXERCITATIONS PHYSICO-ANATOMICAL

By WALT. CHARLTON : *M.D.*

LONDON,

Printed for *Henry Herringman*, and are to be sold at his shop at the *Anchor* in the lower walk in the new *Exchange*. 1659.

Figure 1.1. The title page of what appears to have been the first book on nutrition in English. (The author's name is more commonly spelled "Charleton" but here the "e" is omitted.) (Bancroft Library, University of California, Berkeley)

of carbon dioxide in the air had been discovered, it was thought that rainwater actually provided the "substance" of trees.)

He further thought that, after the food had been worked on in the acid conditions of the stomach, the usable portions were divided up according to their function – the fuel going into the blood for its distribution, and the "restorative principle," or *succus nutritius*, being circulated by another route, which he thought must be the nerve fibers.[14] Fifteen years later, John Mayow expanded this concept in relation to rickets, acknowledging that the idea originally came from the Cambridge professor Francis Glisson. His argument, paraphrased, was as follows: In this disease the child's head appears unusually large but the legs weak and poorly developed. The size of the head shows that there can be nothing wrong with the blood, which is the same wherever it circulates. The only other fibers going to the legs are the nerves. There must therefore be some blockage in the spinal tract which hinders distribution of the nutritious nervous juice from the brain to the lower areas.[15]

Mechanical Models

Charleton did not refer to the qualities of individual foods, nor did he indicate which were good sources of the two classes of nutrients that he discussed. But it was obviously more difficult to understand how plant foods could provide the substances needed for growth or "maintenance" than how animal products such as meat could be reconverted to animal (or human) tissue. The next 70 years were dominated by the successes of Newton and others in explaining phenomena in terms of physics, and mechanics in particular. Not unexpectedly, enthusiasm developed for trying to explain the function of living things as well by a *mechanical physiology* entirely free of chemistry.[16]

It had been suggested as early as 1663 that "the human body itself seems to be but an engine,"[17] and Hermann Boerhaave, the Dutch physician and leading medical teacher of his time, began his textbook for medical students with 99 pages devoted to physics and the admonition that students must familiarize themselves with Newton's *Principia*.[18] Andrew Pitcairn, one of the first lecturers at the Edinburgh Medical School, founded in 1695, concluded that the "animal economy" was no more than a hydraulic system; disease resulted from obstruction of the circulation, and animal heat resulted from friction among the particles in the bloodstream tumbling against the

14. Ibid., pp. 156–62.
15. Mayow (1674), pp. 306–9; Davis (1973), pp. 87–8.
16. Brown (1968), pp. 122–235; Hall (1969), vol. 1, pp. 250–63, 367–408.
17. Boyle (1684).
18. Boerhaave (1719), pp. 1–99.

vessel walls and one another. This also had the effect of mechanically abrading (i.e., grinding down) absorbed nutrients to the appropriate size for deposit in the tissues, whenever there were holes waiting to be filled.[19] The holes had arisen in the first place because of wear on particles in the wall, which then fell out. In the rough and tumble of the bloodstream, these were then finally broken down into pieces small enough to pass out through the pores of the skin as insensible perspiration.

British workers confirmed Santorio's observation of weight loss that could be explained at the time only as being due to insensible perspiration, though not in quite such quantity. Keill reported that with an average intake of 74 oz. food and water, 5 oz. were lost as feces, 38 as urine and 31 by perspiration.[20]

The final conversion of absorbed aliment to usable animal material was thought to result from the combination of the mechanical beat of the heart and the pressure of the air: "In the heart, briskly rubbed together they subtilize and rarefy each other, the more volatile parts grinding and breaking in pieces the grosser and less subtilized; also dulling the edges of the sharper parts."[21] People doing hard physical work – whose hearts were beating more strongly and whose breathing was deeper and more rapid than that of more sedentary people – broke down absorbed food particles more quickly. Delicate food was therefore unsuitable for this class of people, on the assumption that it contained smaller particles, which would quickly be broken down too far and perspired before they could be used to fill in gaps in the tissues. This provided an argument for coarse food being natural and preferable for laborers. By contrast, the richer, and physically less active, classes did not have the digestive vigor to utilize such materials, and actually needed more refined and dainty refreshment.[22]

The same line of argument was used to explain why inactive people needed meat and laborers did not:

> There is a similarity and homogeneity between the muscular flesh of tender, sweet animals and that of the human body; the integral particles of their solids and the component globules of their juices are ready formed... to build up the flesh and furnish out the juices of the latter... with less labour or struggle than those of vegetables in general; as a mason will sooner and more strongly build the walls of a house, who has plain rectangular stones at hand, than one who has rough stones only... which must be first figured or prepared for a solid, durable building.[23]

Enthusiasm for the mechanical approach to medicine was on the wane after 1740, for despite the all-embracing claims for its worth, it had failed

19. Pitcairn (1718), pp. 11, 20, 36–7.
20. Keill (1728), p. 332.
21. Crawford (1724), p. 65.
22. Ibid., pp. 174, 333; Arbuthnot (1731), p. 35.
23. Cheyne (1740), p. xiv.

to produce more effective treatments and was subjected to ridicule. John Arbuthnot wrote in *The Memoirs of Scriblerus:*

It is well known to anatomists that the brain is a congerie of... canals of great length variously intorted and wound up together.... Simple ideas are produced by the motion of the spirits in one simple canal; when two canals merge they make what we call a proposition, and when two empty into a third they make a syllogism. ... That some people think so perversely proceeds from the bad configuration of these canals.[24]

A direct attack on the mechanical system came from the French scholar François Quesnay:

The idea, which many people hold, that digestion consists of mechanical pulverization or grinding is a fiction devoid of any probability. It bears no relation to the character of the digestive organ or of the foods which are digested there. The stomach is a soft and supple pouch which can only be gently agitated.... Also, foods are not typically friable, but doughy or cartilaginous, and usually also soaking in drink in the stomach.[25]

Another of the basic assumptions of the system had been that animal heat could be explained by friction among particles circulating in the blood. This was challenged in the *Proceedings of the Royal Society* for 1745 by a writer who said that, just as in the polishing of marble, greasing or watering was used to prevent friction, so would the fluids of the blood prevent heat from being produced by its circulation.[26] In his opinion, heat came from the fermentation of food and from animal tissues coming into contact with air.[27]

Digestion by Fermentation

In fact, the idea that something similar to fermentation occurred during the digestion of food, or its "concoction" as it was called, was not a new one. These were changes that were easily observed outside the body when foods were kept moist and warm and lost most of their characteristic differences.

The Belgian scholar Jean-Baptiste Van Helmont (1577–1644) had argued that the Ancients' explanation of "vital heat" causing digestion was inadequate. Cold-blooded creatures like fish were as efficient as hot-blooded ones at digestion. There must be some chemical processes at work, and these he attributed to specific "ferments." In the case of the stomach an acid ferment was involved, but one with properties not possessed by simple

24. Aitken (1892), pp. 352–3.
25. Quesnay (1747), Vol. 3, p. 14.
26. Mortimer (1745), pp. 473–4.
27. Ibid., 477.

acids such as vinegar or oil of vitriol (sulfuric acid). The direction of fermentation was also different from that observed to occur spontaneously outside the body.[28] Thus meat left in a warm place undergoes putrefaction, giving off a foul odor, but digestion of meat in the stomach follows a different course.

Thomas Willis (1621–75), a leading medical writer of his time and an Oxford professor before becoming a successful London physician, strongly supported the fermentation concept and considered that it operated throughout the body tissues.[29] He referred to the use of old, sour dough as a starter for making a new batch of bread and suggested, by analogy, that old, "perfected chyle" lodged in the folds of the stomach wall might have the same action in speeding the fermentation of food as it arrived in the stomach.[30]

In some ways it made very little difference whether the digestion of food was considered a mechanical principle or one of fermentation. In either case, the particles present were thought of as being jumbled about and rearranged to give products with different properties. However, later writers did not subscribe to Van Helmont's belief that the fermentation in the stomach was of a particular kind; they assumed that the changes in the digestive system were the same as those seen in the breakdown of materials outside the body. This in turn led to some novel ideas about the principles of good nutrition.

Acid–Alkali Balance

The spontaneous fermentation of grains, flours and starchy or sugary foods such as potatoes or grapes resulted in their becoming acidic. Vinegar production was a well-known example of this. Animal products, however, underwent quite different changes and became alkaline. Robert Boyle, an aristocrat and one of the leading members of the Royal Society from its foundation in 1662, was a pioneer in the attempt to dissociate chemistry from alchemy and to establish it on a rational, scientific basis, and he took an interest in the subject. A material, or vapor, was alkaline, according to Boyle, if it turned a paper soaked in syrup of violets from blue to green; the use of litmus paper, going from red to blue with alkali, was a later development. In 1684 he wrote that putrefied blood and strongly heated hartshorn (deer antlers) each gave off a volatile alkali, which, to him, was identical with that of smelling salts (ammonium carbonate). However, he added that society ladies who visited his laboratory told him that, while the

28. Helmont (1662), pp. 115, 208; Pagel (1956), pp. 524–6; Mendelsohn (1964), pp. 41–3.
29. Davis (1973), pp. 82–6.
30. Willis (1684), p. 12.

smell of the salts was pleasant and refreshing, the smell of the others was disgusting.[31]

François de le Boe (1614–72), who latinized his name to Sylvius, was an influential professor in Holland who argued for the chemical nature of the digestive process. He believed that good health depended on acid and alkaline materials in the body being in balance and thus neutralizing each other. An excess of either caused a condition of *acridity,* which irritated the tissues and was the basic cause of most diseases.[32] From this it seemed to follow that, if people were to remain healthy, they needed a balance of the two types of food that went respectively alkaline and acid when decomposed in the body. This seemed to explain the value of such traditional dietary combinations as bread, which was acescent (acid-producing), and meat, which was alkalescent. A diet tipped slightly to the alkalescent side could be balanced by the addition of vinegar. Scurvy was considered to result from an alkaline acridity, and ships' surgeons were provided with sulfuric acid as a corrective when naturally "sharp" (i.e., acidic) fruits like lemons were not available.[33]

It soon became clear to people pursuing the matter that not all plant materials were acescent. Many types of green leafy materials gave off a volatile alkali either when left to go putrid or when distilled over a fire when still fresh. To the surprise of some, even the leaves of the plants valued as antiscorbutics (e.g., watercress and scurvy grass) proved to be alkalescent.[34] A committee of the Royal Academy of Science in France was set up for the systematic study of the chemical characteristics of different plants.[35] One finding was that mushrooms, either distilled or fermented, gave off so much volatile alkali that "if one did not know better, one would think them to be of animal origin."[36]

Herman Boerhaave, who was, as mentioned before, the most influential teacher of medical theory at the beginning of the eighteenth century, did not agree with all of Sylvius's ideas about acids and alkalies in the digestive processes.[37] However, he pointed out that, in view of animals' ability to live entirely on acescent vegetable foods, the predominantly acid chyle that they produced must gradually be changed to animal tissues with their alkalescent qualities. Therefore, the process of *animalization* of plant foods, as he called it, was one of making them more potentially alkaline. He thought

31. Boyle (1684), pp. 116, 121–2.
32. Boas (1956), pp. 14–15.
33. Boerhaave (1715), pp. 313–21; Arbuthnot (1731), pp. 176–7, 181; Carpenter (1986a), pp. 56–7.
34. Homberg (1702), p. 51; Boerhaave (1715), pp. 19–20.
35. Dodart (1731); Holmes (1971), p. 136; Stroup (1990), pp. 89–100.
36. Lemery (1721), pp. 25, 30.
37. Jevons (1962).

that the heat of the body fostered these chemical changes. Again, he argued that laboring peasants with their more vigorous physique were more capable of this than the physically inactive. Bread and vegetables were therefore suitable for peasants, while the rest of society needed inherently alkalescent foods, like meat and eggs.[38]

The Discovery of Wheat Gluten

The next important finding was made by an Italian scholar who had also combined the study of chemistry and medicine. Jacopo Beccari (1682–1766) spent his whole working life in Bologna at the university and academy. The latter was a novel institution supplied with scientific equipment that the university departments of that period did not have.[39]

In the present context Beccari is remembered for a paper published in 1745 (though, in fact, it was an extended minute of a lecture given in 1728).[40] He argued that "our bodies must, presumably, be composed of the same substances which serve as our nourishment,"[41] yet the obvious properties of wheat flour, like those of other plant foods, appeared quite different and almost opposite to the basically gluey nature of muscle, blood and egg white. His new finding was that, in fact, wheat flour did contain some of this "animal glue." It was normally hidden by the larger quantities of other constituents. He had separated it by first sieving coarsely ground wheat so as to remove the branny particles. Then he added a little water to the flour and kneaded a ball of dough (as in bread making, but without yeast), then continued to knead it under water until all the white floury particles in it (i.e., the starch) had dispersed. The residue was a soft, elastic, glutinous mass insoluble in water. When heated strongly, or kept moist and warm for several days so that it began to putrefy, it gave off a "volatile alkali" with a urine-like smell. This was exactly what happened with animal tissues, but it was quite different from the behavior of whole wheat, or of the wheat starch, which went acidic with a wine-like fermentation if left warm and wet.

Beccari believed, therefore, that he had demonstrated that this vegetable food contained a proportion of what could be called "animal substance" and that this provided an explanation for the reputation of bread as being highly nutritious. He was puzzled, however, at being unable to find similar

38. Boerhaave (1719).
39. Heilbron (1991), pp. 62–3.
40. Beach (1961).
41. Beccari (1745). An English translation has been published by Bailey (1941), but translations of this sentence differ greatly. The original Latin is "Nam si corpus tantum spectemus, immortalemque ac divinum animum excipiamus, quid aliud sumus, nisi id ipsum, unde alimur?"

glutinous material after washing out the floury particles from either barley or beans.

Two historians of science have criticized the priority accorded to Beccari in recent years as the "discoverer of vegetable protein."[42] They point out that many workers had already (i.e., before 1728) shown that some vegetable materials gave off volatile alkali when distilled or allowed to ferment. Indeed, some species had been described as "animal plants" on this basis, and this was also the crucial test used by Beccari to identify gluten as an "animal" substance.

We have already referred to these findings, as well as to Boerhaave's conclusion that animal bodies must have the power to turn acescent foods into alkalescent tissues, since some animals and birds can live almost entirely on grain. We do not, unfortunately, have Beccari's full lecture in his own words, but it seems likely from his introductory point, namely, that "our bodies must, presumably, be composed of the same substances which serve as our nourishment," that he doubted the need for such drastic changes. The expansion of this point would be: Within this food (i.e., bread), traditionally classed as acescent, there is a portion that is already "animal" in nature; therefore, this fraction could be the part used to repair, or replace, body tissue while the remaining, starchy part is either got rid of or, possibly, "form[s] oil for the lamp," to use Charleton's phrase.

Another interesting aspect of Beccari's work was that he had shown the presence of gluten, with its physical, elastic properties similar to those of coagulated animal tissues, by the very mildest of procedures – that is, washing in cold water. There was a concern at this time to develop new and less destructive methods of analysis using solvents,[43] because it was feared that the older methods of analysis by distillation or fermentation might have yielded products that were not present in the starting materials.[44] Thus volatile alkali was produced by distilling materials that were not initially alkaline. Certainly the repeated references to Beccari's work in the second half of the eighteenth century point to the importance attached to it at that time.[45]

A Physiologist's Perspective

Albrecht von Haller, still regarded as one of the most brilliant physiologists of the mid-eighteenth century, sets out interesting views of digestion and

42. Jevons (1963); Goodman (1971).
43. Venel (1755), p. 319.
44. Holmes (1971), pp. 138–9.
45. Van-Bochaute (1786), p. 114; Fourcroy (1789b), pp. 252–3.

nutrition in his general textbook of physiology. He followed earlier writers in thinking of the living body as in a state of continual turnover: "Like the timbers of a ship, it is repaired every day during life, till at length not two jots of the old or first material remain – we may venture to say that in three years the change is universal."[46] "Even bones have a slow dissolution and a perpetual renovation.... The whole living body is in a perpetual state of fluxion, consumption and renovation."[47]

He says that the blood contains "dissimilar parts of various natures."[48] This simple statement can easily be overlooked, but it is important because so many others had thought of it as a single "thing," with regards to both its manufacture and its conversion into tissue. He goes on to say that "the red parts of the blood (i.e. the globules) seem to be chiefly used to produce heat.... The hardening serum is more especially designed for the secretions and nutrition of the parts."[49] These are made up of chalky particles cemented together by glue or jelly.[50] Elsewhere Haller suggests that the various uses of fat are: as a lubricant for muscles moving against each other, "giving a comely form and cushioning against knocks," and having a primary role in the formation of bile.[51]

Haller believed that the action of the stomach was mainly mechanical, with food being rapidly moistened and gradually "opened up" to allow extracted juices to diffuse out and move into the blood.[52] Only the flesh of animals contains "the gelatinous lymph, ready prepared for the recruit both of our solids and fluids..., and from our canine teeth and the shortness of our cecum it appears that animal food is a necessary part of our nourishment." However, vegetable foods do have "a small portion of jelly drawn from their farinaceous [starchy] parts which, after many repeated circulations, is converted into the nature of our indigenous juices." Vegetables were also thought to serve a kind of "diluting" function: "They are necessary to avoid over-repletion with blood... [which is said] to breed a fierce or savage temper... only to be avoided by a change of diet in which a vegetable acidity abounds. Hence it is that we are furnished with but few canine teeth, and our appetite is strongly for acidulous vegetables."[53]

Yet Haller recognizes that cattle and other species live on nothing but vegetable materials, and he argues that these animals have the power to

46. Haller (1754), Vol. 1, p. 10.
47. Ibid., pp. 195–6.
48. Ibid., p. 136.
49. Ibid., p. 160.
50. Ibid., p. 6.
51. Ibid., p. 21.
52. Ibid., Vol. 2, p. 157.
53. Ibid., pp. 122–3.

convert the plant materials that become viscid (sticky, viscous) when mixed with water into a "coagulable lymph."[54]

Other physicians at this time held the opinion that any viscous solution, whether from cooked starch or gums, was automatically nutritious and could be turned into blood, another sticky fluid. A ship's surgeon facing an outbreak of scurvy at sea had said that "the blood was so thin as to be unfit for circulation." From this it seemed to him that foods of a glutinous nature, such as salt fish and bread, were the most suitable for long voyages. And in 1777 James Lind was to write that an ounce of powdered salep (dried orchid root) and another of "portable soup" (dried cakes of boiled-down cattle offal, with seasoning) would make two quarts of a thick jelly and serve as complete sustenance for a man for a day.[55] The modern nutritionist would, of course, say that these 2 oz. of dry matter could provide only 10% of a man's daily needs, but we will see that exaggerated ideas of the value of soup were to persist for at least another century.

Potatoes as a Substitute for Grains?

The next personality to contribute to our subject, a generation later, is the French food scientist Antoine Parmentier. Born in 1737, he trained as a pharmacist and eventually became chief pharmacist to the French army. The records show that he had a long-term concern for the welfare of ordinary people and the effects of periodic, local famines that afflicted the French countryside during this period.[56] While serving with the army in Germany during the Seven Years' War, he noticed the importance of potatoes in the peasant diets in the areas where he was stationed. This crop was still almost unknown in France, although it had been brought to Europe from the New World at least 200 years earlier. Back in France, he demonstrated that potatoes could be grown successfully in sandy soils that did not support grain production. However, people were generally skeptical as to whether potatoes were safe and "meant to be eaten" since they were not mentioned in the Bible. Parmentier sought official support for his "Grow more potatoes" campaign, and was successful, in part, because of a seemingly irrelevant gesture by the king and Marie Antoinette, who accepted from Parmentier some flowers from a potato plant and wore them as corsages for a day. This was regarded as symbolic of royal support for the campaign.[57]

In his writings Parmentier was concerned to demonstrate that the potato

54. Ibid., p. 199.
55. Carpenter (1986a), pp. 49, 67.
56. Berman (1974).
57. *Grande Encyclopédie* (1899), Vol. 25, pp. 1177–8.

could serve as a substitute for wheat flour, and he submitted the winning essay to a French provincial academy of science that had offered a prize for the best advice on substitute foods that could be used when wheat was scarce.[58] There was no problem over the quantitative yield of dry matter from the new crop – it was at least equal to, and usually greater than, that from wheat. However, with regard to quality, Parmentier had to take into account Beccari's demonstration of the gluten, with its animal character, in wheat. He also referred in his writing to a thesis by Johanne Kesselmeyer, published in Strasburg in 1759. This reported a series of experiments with wheat gluten and the author's conclusion that its properties made it virtually identical with the material of blood serum, which explained why wheat was so nourishing.[59] Parmentier was unable to extract a similar, glutinous fraction from potatoes, though he, too, did many experiments with wheat gluten, demonstrating that it was the factor that allowed fermented dough to rise during baking and then not to collapse.[60] In addition, and this is an interesting general point, he accepted that the fact that both insects and birds seemed particularly attracted to wheat must be evidence of its general nutritional value.[61]

But if wheat consisted essentially of a large proportion of starch and a small proportion of gluten, and only the latter was truly nutritive, what was the nutritive value of potato, in which only starch had been identified? It seemed to follow that it had to be zero. Parmentier considered the problem and concluded that, since people obviously could thrive on a mainly potato diet, Kesselmeyer and Beccari must be mistaken in their conclusions; there was, in any case, too little of the gluten for it to be the main nutritive principle of wheat. Given the facts, as he knew them, he could only conclude that the true nutritive material in both foods was the starch.[62]

In so many puzzling questions that nutritionists have had to consider in the past, the apparently simple choice of "either this *or* that" has turned out to be spurious. As a French scholar has written in this century: "Logic is an unreliable instrument for the discovery of truth, for its use implies knowledge of all the essential components of an argument – in most cases an unjustified assumption."[63]

In the present example, vital additional information was published in the very same year that Parmentier's book appeared – 1773. The author was H. M. Rouelle, born in 1718 and the younger, and by then the survivor,

58. Balland (1902), pp. 1–20.
59. Kesselmeyer (1759).
60. Parmentier (1773), p. xx.
61. Ibid., p. 244.
62. Ibid., pp. 237–9.
63. Dubos (1950), p. 131.

of two brothers who were both famous French chemists of this period.[64] He described in two papers how he had discovered soluble "animal substance" or, literally, "animal–vegetable glutinous material" in juices squeezed out of fresh green plants as different as sorrel, hemlock and chervil.[65] In his second paper he gave a full account of this procedure using hemlock taken just at the point of flowering.[66] The plant material was carefully crushed in a mortar and then pressed. The juice was filtered through a cloth and then heated to a point at which he could just bear to immerse his fingers in it for a few minutes. (He had no thermometer.) A green precipitate formed, and this was retained on a filter cloth, transferred to a glazed pot and stirred with 8 to 9 pints of added water. After 24 hours, the water was poured off and the process repeated twice more. The remaining precipitate was transferred to another filter cloth and blotted to take up as much water as possible. It was then subjected to further extractions, this time with the same quantity of hot alcohol in a double boiler, until all the green coloring material had been removed. The alcohol was "spirit of wine," containing perhaps 80% by weight of ethanol, in view of the crude distilling apparatus available at the time.

The residue was a dirty whitish gray, and darkened further on being dried. When it was placed in a retort and heated, the distillate contained an "alkaline volatile spirit," that is, ammonia, and finally a "concrete volatile alkali" (crystals of ammonium carbonate?). The sequence of observations and the appearance of the residue were exactly the same as that obtained with either wheat gluten or the caseous curd of milk and was proof for Rouelle that he had obtained another glutinous or vegeto-animal substance.[67]

Summing Up

In 1773 chemistry was about to be revolutionized by some sensational developments. This is therefore a good place to summarize the knowledge gained and the changes in ideas that occurred over the 120 years from the 1650s to the 1770s.

Constant throughout the period was the assumption that respiration served the purposes of cooling the heart and, through air pressure, aiding physically in the digestion, or concoction, of foods for their conversion into blood. Whether this conversion was achieved purely mechanically or whether it was the result of specific types of fermentation was a subject of

64. Rappaport (1960); Partington (1962), Vol. 2, p. 76.
65. Rouelle (1773a), pp. 264–5.
66. Rouelle (1773b).
67. Ibid., pp. 61–6.

debate, with the evidence for neither view strong enough to discredit the other completely.[68] In either case it was agreed that the main function of nutrition was to replace rapidly turning over tissues and that the degraded tissues broke down further into particles small enough to diffuse through the skin in the form of insensible perspiration.

Toward the end of the period, it was suggested that only the fraction of vegetable foods that had alkalescent properties was truly nutritive – that is, capable of replacing worn-out animal tissue. At first, such a fraction was found only in grains of wheat, but at the very end of the period, similar fractions were found in green plants as well. However, it was thought unlikely that these alkalescent fractions could be the only nutritive portions since they made up such a small percentage of the whole. Was everything else useless?

At the beginning of the period, Walter Charleton had referred to a portion of ingested food being needed as oil or fuel for the "vital flame burning within us." However, with the growth of the mechanical philosophy, animal heat was assumed to come from friction among particles in the bloodstream. Similarly, for those who considered fermentation to be the main mechanism of digestion, the violent reactions in progress were thought to be the source of animal heat. The idea of something analogous to a flame requiring a specific fuel dropped from sight. It seemed to make more sense that the body was capable of turning every kind of digestible food into animal tissue by one means or another.

The developments described in the next chapter were to throw entirely new light on all these matters in less than a generation.

68. Holmes (1975), p. 138.

2 Nutrition in the Light of the New Chemistry, 1773–1839

THE PERIOD FROM 1770 to 1790 saw the development of a "new chemistry" that changed the ways in which the utilization of foods could be studied. The story has been told many times, at length, elsewhere.

It began with the studies of combustion in Britain over the preceding hundred years. Why did the flame of a candle go out after a period of burning in an enclosed space? Why did a mouse die quite quickly when put into the space in which a candle had been burning, whereas a green plant seemed to thrive in it? "Was there only one kind of air, or was ordinary air a mixture?" Such questions were puzzled over in the Royal Society and elsewhere. They were also thought to have medical importance, since ill health was often associated in people's minds with bad air and lack of ventilation in slums and prisons, or below deck in ships of war where sailors developed scurvy. There was also a philosophical problem: Why was it that, despite the general principle of attraction between individual particles, as expressed in the law of gravity, the particles in gases seemed to repel one another? And yet a material like chalk seemed to contain a highly compressed gas that expelled itself into a hugely greater volume when the chalk was heated and converted to lime.[1]

Antoine Lavoisier

The person who did the most to construct a rational scheme of chemistry that has survived to this day was the French scientist Antoine-Laurent Lavoisier. To quote a specialist: "The chemical Revolution was the work of many hands... [it] was prepared on French soil with materials quite conspicuously of British origin."[2] It was completed under the storm clouds of a political revolution.

1. A short introduction to these findings and their relation to Lavoisier's work has been given by McKie (1965).
2. Guerlac (1961), p. xiv.

Lavoisier was a complex person, and his work as well as his personal characteristics have been analyzed in great detail, as befits someone who made such important contributions to science.[3] He came from a family whose members had in successive generations risen in status from postilion to postmaster to lawyer. He himself was brought up in some luxury and given every opportunity for study. He was expected to be a lawyer and to hold official positions. After studying hard and qualifying for the practice of law, he was already wealthy enough from his inheritance to be able to work, as an amateur, in a geological field survey with a powerful patron. He interested himself mainly in mineralogy.

In 1771, at the age of 27, he married a girl only 14 years old, but intelligent and well educated, as well as an heiress. Through this new family connection he became a member of the national tax-gathering syndicate (la Ferme-Générale), which collected taxes in return for prearranged payments to the state. The syndicate also had a monopoly on the sale of tobacco, and Lavoisier was assigned to the supervision of that part of its activities. Apart from his work for the syndicate, his substantial income allowed him to give the remainder of his time to his own scientific interests.

In 1772 he became a member of an official committee of the Royal Academy of Sciences, examining the claims in a paper regarding the complete disappearance of diamonds when they were heated strongly in a furnace. It was regarded as extremely mysterious for a gemstone to behave in this way. Was it a case of combustion? At that time, combustion was viewed as a material losing something, a "matter of fire," that could be seen leaving the material in the flame and in some way combining with the air, until the air could take no more.

Lavoisier began to look critically at this view of combustion and to make crucial observations for himself, doing quantitative weighing experiments using the best possible balances. His first experiments confirmed that both sulfur and phosphorus gained weight when they burned. In later work he showed that a given weight of "inflammable gas (hydrogen)," obtained from the reaction of metals with acids, yielded a greater weight of water when it was burned in air. This dealt a decisive blow to the idea that water was an element and made it easier to believe that "air" was not a single element either.

Finally, in 1789 Lavoisier assembled the new scheme in his *Traité élémentaire de chimie*, which was immediately translated and published in Edinburgh as *Elements of Chemistry*. Part of one table from the English translation is reproduced here (Figure 2.1). What he called "simple sub-

3. Detailed studies of Lavoisier's chemical work have been made by Guerlac (1961) and Holmes (1985). Recommended biographies, including parallel studies of his British contemporaries, are those of Aykroyd (1935) and Davis (1966).

TABLE OF SIMPLE SUBSTANCES.

Simple ſubſtances belonging to all the kingdoms of nature, which may be conſidered as the elements of bodies.

New Names.	*Correſpondent old Names.*
Light	Light.
Caloric	Heat. Principle or element of heat. Fire. Igneous fluid. Matter of fire and of heat.
Oxygen	Dephlogiſticated air. Empyreal air. Vital air, or Baſe of vital air.
Azote	Phlogiſticated air or gas. Mephitis, or its baſe.
Hydrogen	Inflammable air or gas, or the baſe of inflammable air.

Oxydable and Acidifiable ſimple Subſtances not Metallic.

Sulphur	
Phoſphorus	
Charcoal	
Muriatic radical	Still unknown.
Fluoric radical	Still unknown.
Boracic radical	Still unknown.

Oxydable and Acidifiable ſimple Metallic Bodies.

Antimony	Mercury.
Arſenic	Molybdena.
Biſmuth	Nickel.
Cobalt	Platina.
Copper	Silver.
Gold	Tin.
Iron	Tungſtein.
Lead	Zinc.
Manganeſe	

Figure 2.1. Lavoisier's list of elementary bodies from his *Elements of Chemistry* (1790).

stances, that have not yet been decomposed," we would call "elements." The major elements of living bodies – carbon, hydrogen, oxygen and nitrogen – are all there. Lavoisier gave "hydrogen" its name because it could be converted to (or was a source of) water when burned. ("Hydro" is the Greek root for "water" and "gen" gives the idea of source, as in "genesis" or "generate.") He coined the term "oxygen" (or "acid former") because of his experience with the combustion products of carbon, sulfur and phosphorus. In each case the oxides dissolved in water to form acids. He con-

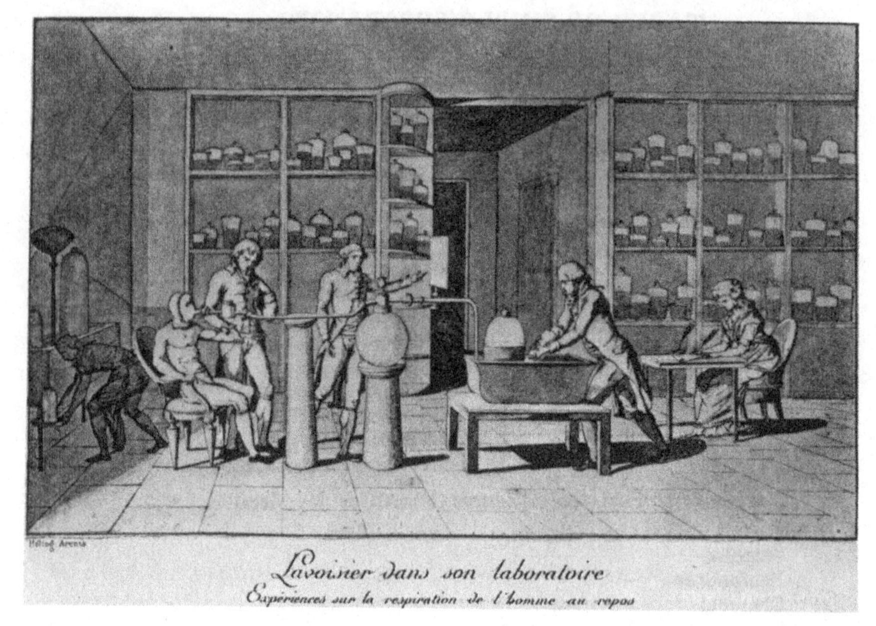

Lavoisier dans son laboratoire
Expériences sur la respiration de l'homme au repos

Figure 2.2. Lavoisier experimenting on human respiration with Madame Lavoisier taking notes. (Wellcome Institute, London)

sidered using the word "nitrogen" (or "nitigen") but eventually chose (and the French have retained) "azote," meaning the gas that does not support animal life.[4] ("Nitrogen" was the name adopted in English since, when sparked with excess oxygen, this element gives a gas that dissolves in water to yield nitric acid, well known at that time.)

Lavoisier also conducted physiological experiments with a younger colleague, Armand Seguin. They measured the uptake of oxygen and production of "fixed air," or "carbonic acid" (as he called it), during the respiration of both humans and guinea pigs, and concluded that the ultimate source of the carbon expired was the food eaten. However, they believed that atmospheric nitrogen was neither consumed nor excreted through the lungs. Lavoisier also found that the rate of combustion greatly increased when the subject was doing physical work (Figure 2.2). From experiments in a calorimeter with guinea pigs, he concluded that animal heat could be explained as coming from the oxidation of carbon and hydrogen in our tissues (or ultimately from our food).[5]

In their report of these experiments, Lavoisier and Seguin pointed out

4. Lavoisier (1790), pp. 48–53.
5. Seguin and Lavoisier (1789).

that respiration now had to be considered in terms of using up the body's tissues. It was ironic, therefore, that the rich person who did no physical work needed less food than the laborer, although it was the latter whose ability to purchase food was so limited. They ended by expressing the hope that the reforms then in progress would correct this imbalance.[6] Lavoisier was obviously taking the opportunity to show that, despite his own wealth and membership in the tax syndicate, his sympathies were with the new regime and the common people. But it is also true that his new findings had far-reaching implications for the understanding of nutritional needs, and these will be considered in due course.

One wonders what Lavoisier would have done next. But in November 1793, although he had been heavily engaged in work for the revolutionary government, in the organization of gunpowder supplies and an improved system of weights and measures, and was also a commissioner of the National Treasury, he was arrested. He was imprisoned with other former members of the Ferme-Générale, who included his father-in-law. They were ordered to complete an audit of their accounts up to 1791, when the syndicate had been suppressed.[7] The government hoped to confiscate large, fraudulent profits, but apparently there had been no excess profits. A cynic has said that the syndicate's legal profits were large enough to make fraud unnecessary. Its accusers then relied instead on charges that the tobacco it sold had been adulterated with water in order to increase its weight. Lavoisier tried to explain that this was a normal part of the "curing" process, but to no effect. It seems likely that the syndicate had been tempted to add more water than was really necessary, despite Lavoisier's protests at the time. And, of course, if the man in the street had had difficulty over the years in keeping the tobacco in his pipe lit, there was an obvious target for his abuse – the greed of the tax collectors – and now there was an opportunity for revenge.

In May 1794 the syndicate members were brought to a hurried trial, and all 28 were guillotined the same afternoon. Lavoisier is believed to have asked for a short reprieve in order to carry out one last experiment on transpiration, but the presiding judge is said to have replied: "The Republic has no need of savants [scientists *or* scholars]." Later generations have expressed surprise that leading scientists among Lavoisier's contemporaries did not do more to try to save him. During the Reign of Terror at this time it was obviously dangerous to appear to be in sympathy with someone accused of being a traitor, but it has also been the general conclusion that Lavoisier had made enemies and that others were jealous of his wealth and scientific success.

6. Ibid., pp. 578–9.
7. Aykroyd (1935), pp. 155–92.

IDENTIFICATION OF THE VOLATILE ALKALI

We return to the subject of the alkalescent "animal substances" in food. Lavoisier did no work on them himself, but in 1778 he wrote a report for the Academy of Sciences on three memoirs submitted by Parmentier on methods of measuring the quality of wheat and flour, and how to make best use of samples that were beginning to deteriorate. We discussed in Chapter 1 Parmentier's view that starch was really the nutritive factor in wheat, and this was repeated in the memoirs. Lavoisier wrote: "It has to be accepted that other grains such as barley and oats, which do not contain gluten, are also nutritious. Starch must therefore be nutritious, but Parmentier seems to us to go too far in singling out starch as the only nutritive element in wheat. It seems obvious that gluten, with its chemical properties, is more 'animalized' and therefore eminently nutritive."[8]

The first advance came in 1785 when Claude Berthollet told the Academy of Sciences in Paris that he had discovered the composition of the volatile alkali associated with animal substances.[9] Translating his paper into more modern chemical terms, we may describe three of the results as follows:

1. Reacting "volatile alkali" (ammonia gas) with chlorine resulted in the production of nitrogen gas.
2. Reacting ammonia gas with hot metallic oxides resulted in the production of hydrogen.
3. Electric sparking in ammonia gas gave a greatly increased volume of gas. This was then burned in oxygen with the production of water (proving the presence of hydrogen) and leaving an unreactive gas (assumed to be nitrogen).

Berthollet estimated that the compound contained one-sixth of its weight as hydrogen. (This is close to the modern value of 1 part in 5.6.)

Lavoisier himself did further work on the composition of isolated plant fractions and concluded that sugars, gums and starches consisted of combinations of carbon, hydrogen and oxygen only.[10] By contrast, animal tissues, in the light of Berthollet's finding, obviously contained nitrogen as well.

Berthollet argued that ammonia was not present as such in healthy tissue and that it was formed by a combination of nitrogen and hydrogen during the process of either distillation or fermentation.[11] He was impressed by repeated observations that treating animal substances with nitric acid resulted in the immediate release of nitrogen gas. (Presumably his "nitric acid"

8. Lavoisier (1778).
9. Berthollet (1785a, b).
10. Lavoisier (1790), pp. 117, 142–3.
11. Berthollet (1786), p. 273.

was impure and included a proportion of nitrous acid that would really have been responsible for this reaction.) There was no such reaction when he added nitric acid to ammonium salts.

How Many Animal Substances?

Antoine Fourcroy, another French chemist interested in applying his science to medicine, read a paper to the Royal Academy of Medicine in August 1788.[12] He supported and extended Berthollet's findings of the presence of nitrogen in isolated animal substances. He said that there was a perfect correlation between the quantity of nitrogen produced by the action of nitric acid and the quantity of ammonia that distilled over when the products were heated.[13]

Fourcroy also said that there were three classes of animal matter. The first was *gelatin*, the jelly that one could extract (by boiling) from skin, tendons, membranes, and other such tissues. The second was *albumin*, which was soluble in water but was precipitated by heat, acids or alcohols; it was found in the white of egg, blood serum and milk casein, and it contained a higher proportion of nitrogen than gelatin did. The third was *fibrin*, which formed clots in blood and which "existed so abundantly in muscles." It contained the highest level of nitrogen.[14] He was surprised to find little more nitrogen in the flesh of carnivorous animals than in that of fruit eaters. Fish were also similar to land animals in their nitrogen content.

In a second paper, published in 1789, Fourcroy reported his further analyses of plants. He confirmed Rouelle's work (described at the end of Chapter 1), though he concluded that the "animal substance" prepared from the juice of leaves was not a form of gluten but had all the properties of animal albumin, including coagulation upon heating and redissolution by alkalis. He found similar material in the roots of plants and in wheat. He made the last discovery by following Beccari's original procedure for separating the gluten from wheat flour by washing the dough in cold water. He then strained the washings to separate out the suspended starch particles. When the clear filtrate was heated, an albuminous precipitate appeared. Fourcroy concluded that wheat contained *two* "animal substances" – albumin, which was analogous to that found in blood serum, and gluten, which was analogous to fibrin.[15]

Fourcroy's final finding was that acescent plants, and acidic fruits in particular, contained absolutely no albumin. Instead, they consistently

12. Smeaton (1962), pp. 19–37.
13. Fourcroy (1789a), p. 43.
14. Ibid., pp. 41–2.
15. Fourcroy (1789b), pp. 255–9.

yielded a gelatin-like material. He suggested that this vegetable jelly was the product of albumin reacting with acid. Then, since all acids were believed to contain oxygen, he suggested that gelatin differed from albumin only in containing a higher proportion of oxygen.[16] Fourcroy was to be vigorously attacked some 12 years later for some of these claims. In particular, it was said that his "vegetable jelly" was well known to be only an aromatic resin, soluble in alcohol and with none of the special properties of a glutinous substance.[17] Certainly Fourcroy made no mention of the "jelly" yielding either nitrogen or ammonia, and one wonders, with hindsight, if he had isolated pectin. A more detailed study of vegetable glue confirmed that it was not gluten-like or an "animal substance."[18]

Like so many of the leading chemists of this period, Jacob Berzelius of Sweden took an interest in "animal chemistry." From a study of the composition of blood he concluded that fibrin, albumin and the coloring matter (i.e., hemoglobin) resembled one another so intimately that they could be considered modifications of a single substance. He called them "albuminous" and concluded that secretions such as the casein of milk were essentially "modified albumin."[19] This was not a novel conclusion by 1813, but carried weight because of Berzelius's reputation for careful, accurate work. In passing, it has been said that "he left school with the usual report on pupils of genius: that there was no hope for him."[20] We will return to his fundamental contributions in the next chapter.

A Nutritional Reassessment

Already in 1786 Van-Bochaute had developed the argument that there was one basic "animal substance" and that from Beccari's and then Rouelle's work it could be concluded that it was manufactured only by the vegetable kingdom.[21] Tactfully, he suggested that Rouelle must have seen the implications of his finding for animal physiology, but had been too modest to make a show of them.

He made fun of the physiologists who tried to persuade their readers that digested foods, whether they consisted of starch, sugars or fat could, almost in a moment, be metamorphosed into "animal substance."

> Thanks to modern chemistry which can throw new light on these obscure sciences. ... we can now understand the process of digestion; we see that it really consists of

16. Ibid., pp. 260–1.
17. Proust (1802), pp. 109–10.
18. Bouillon-Lagrange (1805), p. 36.
19. Berzelius (1813), pp. 71, 115–6.
20. Partington (1964), Vol. 4, p. 142.
21. Van-Bochaute (1786).

a solubilisation of the food materials accompanied by slight decompositions or fermentations, which allow the animal substance, already formed by the vegetable economy, to separate itself from the other components . . . and to pass into the blood circulation by resorption from the surrounding veins.[22]

Van-Bochaute went on to say that this absorbed "animal substance" had different uses, forming the casein of milk for the nutrition (i.e., building new tissues) of the young and for the nutrition (i.e., restoration) of tissues throughout the adult body. The other components absorbed from the gut were not wasted. The "buttery part," which was a vegetable oil, furnished the fat laid down by animals; other portions were used, for example, as antiseptics that corrected the "corruption leading to scurvy." All the evidence had led him to believe that animal matter was manufactured solely by plants. They were particularly suited for it because of the slowness of their chemical processes and their organization, which was designed to utilize the primary elements under the influence of heat, light, water and the minerals of the earth.

Van-Bochaute's paper, published in February 1786, was probably written before Van-Bochaute had read Berthollet's discovery of "animal substance" containing nitrogen. But that discovery, together with Seguin and Lavoisier's 1789 paper in which they said that atmospheric nitrogen took no part in respiration, would only have strengthened his argument. In essence, of course, it was that the building of new tissues and repair of existing tissues was achieved only by absorbed "animal substance" and that the other digested components of foods – sugars, starch, fat and so on – were used, but only for other purposes.

The Concept of Animalization

Van-Bochaute's paper was an exception to the main line of thought after the recognition that animal substance(s) contained nitrogen in greater abundance than plants. Others seemed to have in mind two assumptions. The first was still that the only function of nutrients was to replace worn-out tissues, or to produce new ones, which meant that only nutrients in the form of "animal substance" were of value. The second (already mocked by Van-Bochaute) was that materials of lower nitrogen content could be turned into "animal substance" by being topped up with nitrogen; that is, the character of organic materials was determined entirely by their relative contents of carbon, hydrogen, nitrogen and oxygen. This process was described as "animalization," and the only debate concerned whether the nitrogen came from the air or from organic nitrogen left over when the carbon and hydrogen of some food molecules were oxidized so that the

22. Ibid., p. 113.

residual "organic" nitrogen could be transferred to other materials to raise their nitrogen content.[23] The oxidized material might be decomposing, worn-out tissue rather than newly arrived nutrients, so that the process would not really be as wasteful as it seemed.

Hallé, writing in 1791, developed an argument on the basis of his belief that all the nutritious components of foods (sugar, gum, starch jelly, albumin, casein, fibrin, etc.) when treated with nitric acid gave a residue of oxalic acid. They therefore contained the same "oxide hydro-carboneux" (carbon–hydrogen–oxygen compound) combined with either extra carbon or extra nitrogen. He went on to suggest that, when ingested foods contained this "base substance" with additional carbon, the extra carbon was removed in the digestive tract or in the lungs and skin, where it came into contact with air, and oxidized to form carbonic acid. Nitrogen then took its place.[24]

Medical school teachers in the "physiology" tradition continued to write about nutrition with little regard to the new views of chemists that elements could not be interconverted to one another. A standard history of medicine tells us that "among clinical teachers of the 18th century there is no name more highly esteemed than that of William Cullen (1712–90)."[25] Cullen did, in his final work, make the point that Hallé's gelatinous matter from vegetables was interconvertible with starch and sugar, rather than being albuminous in nature, and also that there was only a little difference between starch and the gums extractable from salep. However, he said that, despite Beccari's finding, he was "still of the opinion that the chief part of the aliment of vegetables is afforded by acid, sugar and oil, which are compounded by the powers of the animal economy."[26]

Benjamin Thompson (1753–1814) was another scientist who ignored the "new chemistry" in his writing on nutrition. He was an American, but one who took the royalist side during the War of Independence and became a colonel in the British army. In 1784 he moved to Bavaria, where he rose to head the army and receive a title as Count Rumford. He was at home in both London and Paris, active in both the Royal Society and the Institut de France, and even married Lavoisier's widow, though they subsequently separated. He had a genius for technological innovation to increase the efficiency, first of an army, and then the operations of ordinary life. He was particularly interested in the conservation of heat and the design of stoves and, for example, is credited with the invention of a drip-type coffeemaker.[27]

He was also concerned with developing cheaper, but still nutritious food for the poor and believed particularly in the value of soups. He accepted

23. Fourcroy (1792), Vol. 2, p. 486; Holmes (1963), pp. 61–5; Florkin (1972), pp. 117–23.
24. Hallé (1791), pp. 162–3.
25. Garrison (1929), pp. 357–8.
26. Cullen (1789), Vol. 1, pp. 169–73.
27. Brown (1976).

the idea discussed in Chapter 1 that all thick soups were equally nutritious. Therefore, since barley flour would thicken three to four times as much soup as the same weight of wheat flour, he believed that it had three to four times the nutritional value, and he put this idea into practice in the provision of soup for the poor in Munich. In an essay written in 1798 he rationalized this principle by suggesting that the discovery of water being a "compound" would lead to further nutritional findings. He said that "our knowledge in regard to the science of nutrition is still very imperfect," but he thought it likely that, if provided in a thickened form, water was capable of decomposition by the body, thus itself serving as a food.[28]

In a book published in Paris in 1806, it was in fact suggested that nitrogen was formed from the partial removal of oxygen from water, and also that carbon, in turn, was formed from nitrogen and hydrogen in one proportion, and ammonia from another proportion of the two.[29] Today, of course, we have abundant evidence that carbon, nitrogen and hydrogen are all elements, whereas ammonia is a compound made from nitrogen and hydrogen. But the proofs for some materials being "elements" were not that strong at the beginning of the nineteenth century, so it is not really surprising that Lavoisier's "new chemistry" was not universally accepted.

The Work of François Magendie

Magendie is an important person in our history – famous for carrying out what now seem extremely simple experiments. Of course, it is with hindsight that the experiments seem obvious and simple. But why had no one previously done them? Was there something special, or different, about Magendie's background? Born in 1783 and brought to Paris at the age of 8, with his father heavily involved in revolutionary politics and his mother dead a year later, he was 10 years old before he learned to read or write. Then he asked to enter elementary school, and at 16 became apprentice to a surgeon friend of his father, for whom he carried out anatomical dissections. At 20 he became a formal medical student. After graduation he practiced surgery but then changed to physiology, becoming a respected teacher and in time the accepted leader of experimental physiology in France.[30]

His first paper that concerns us, "The Nutritive Properties of Substances That Do Not Contain Nitrogen," was read to the Academy of Sciences in 1816.[31] In view of its historic interest, I will set out Magendie's argument

28. Rumford (1800), pp. 194–9, 290–3.
29. Thouvenel (1806), Vol. 3, pp. 26–8.
30. Grmek (1974).
31. Magendie (1816).

at some length. He starts by saying: "Nutrition has often been the subject of conjectures and ingenious hypotheses by physiologists – but our actual knowledge is so insufficient that their only use is to try to satisfy our imagination. If we could arrive at some more exact facts, they would be important for physiology and could well have applications in medicine."

Magendie then reviewed the differing ideas as to the source of the nitrogen in animal tissues. Those who believed that it need not come from food had referred to reports of Arabs who ate nothing but gum when crossing the African desert and of colonial natives who ate nothing but sugar. He was skeptical of these claims and pointed out that people who asserted that herbivores had a nitrogen-free diet were simply wrong: all plant leaves had been found to contain nitrogen.

His own plan was to use dogs – since they, like ourselves, could eat both plant and animal foods – and to feed them on a single food that was accepted as nutritious but free of nitrogen. He gave his first dog an unlimited supply of pure sugar and distilled water. It maintained a reasonably good appetite for about 2 weeks, but was already losing weight and then developed an ulcer on the cornea of each eye. After a month it died. Postmortem examination of the carcass showed a total absence of fat and only one-sixth the normal muscle mass.

Magendie thought it important to verify that these events were the inevitable consequence of living on sugar alone. So he repeated the trial with two more dogs and obtained essentially the same results. He then decided to find out whether the results were caused by a particular property of sugar, and he tested other materials that were generally considered to be nourishing even though they contained no nitrogen. He performed separate feeding experiments with olive oil, gum and then butter. All of the dogs died in less than 40 days with the same characteristics at autopsy, but there was no ulceration in the two dogs receiving olive oil.

Magendie wrote that these facts seemed to throw doubt on the belief that oils, fats, gum and above all sugar were preeminently nutritive, even though they were well absorbed and supported life in dogs for a longer period than if the animals were starved completely. The work also seemed to make it probable, if not certain, that the majority of the nitrogen present in animal tissues came from the food eaten.[32]

He suggested that the results also indicated a promising treatment for people suffering from kidney stones. Most of these stones contain high proportions of uric acid and phosphate. The patients tend to be heavy people who are big eaters of meat, fish and dairy products, so that they have a high nitrogen intake and therefore a high excretion rate. Would it not be reasonable to have them reduce their intake of these foods so as to reduce

32. Ibid., pp. 66–75.

the concentration of uric acid in their urine? Magendie said that his experiences so far had supported the idea, though the change of diet seemed to take a month to have an effect.[33]

One's first reaction to this account is, perhaps, revulsion at the suffering of the dogs. And in his own time Magendie was heavily criticized for his physiological experiments on animals. But we have to remember two things. First, the results with the first animals were so contrary to general expectation that they had to be repeated if they were to be believed. Second, Magendie hoped that his nutritional work would improve human health. Ill health requiring surgery involved acute pain and suffering – there were no anesthetics at the time. Therefore, it must have seemed reasonable to advance knowledge at the cost of some animal suffering.

We must now evaluate the work, and the conclusions, in terms of what was known at the time. Certainly, the results showed clearly that dogs could not live on any one of the test materials alone. The property that Magendie recognized them to have in common was their lack of nitrogen. To rule out the possibility that sugar had some uniquely bad quality, he repeated the test with three other highly regarded foods containing little or no nitrogen. He again found that the dead animals had less than normal nitrogenous material in their urine and feces. But, of course, even if a healthy animal were capable of making nitrogenous compounds from atmospheric nitrogen, this does not mean that a sick animal too weak even to eat would have the same ability. One critic argued that the results showed only that "an animal cannot be supported by a highly concentrated aliment."[34] Presumably he meant that a healthy diet had to include some fiber or ballast, and it is true that Magendie's diets were all lacking in it.

Two modern scholars, making a passing reference to Magendie's work, wrote: "He reported in 1816 that dogs fed a diet completely lacking in nitrogen shortly died, and that the addition to such diets of nitrogen-containing protein materials prevented this early mortality."[35] In fact, they were using their imagination in this description. The really weak point in the work was that Magendie had no positive controls, that is, with an added nitrogenous material such as gluten or albumin. He fell into the same trap as those he had criticized, by saying: "Everyone knows that dogs can live very well on bread alone." Was this a fact any more rigorous than the belief that Arabs could live on nothing but gum? Magendie probably asked himself the same question, or perhaps others put it to him. In any case, the second edition of his textbook of physiology, published in 1831, contains these short statements:

33. Ibid., p. 77.
34. Paris (1826), pp. 96–9.
35. Greenstein and Winitz (1961), Vol. 1, p. 247.

Since the publication of these facts in the former edition of this work, I have been enabled to establish some other very important facts. . . . A dog, eating at discretion pure wheaten bread, and drinking common water, does not live above fifty days. It expires at that period, with all the known marks of final decay recorded above. A dog, eating exclusively of soldier's biscuit, lives very well, and his health is not in any degree impaired. Some dogs, fed exclusively upon cheese, and others upon hard eggs, lived a long time, but they were weak, meager: they lost their hair, and their appearance announced an imperfect nutrition. . . .

[A section on experiments with other species including rabbits and rodents is omitted here.]

The most general, and the most important consequence to be deduced from these facts, which it would be worth while to follow up and investigate anew, is that diversity and multiplicity of aliments is an important rule of hygiene; which is, moreover, indicated to us by our instincts, and by the variation induced by the seasons over all nature, particularly in the species of alimentary substances.[36]

So here were upsetting results: Dogs were receiving nitrogenous foods and still not thriving. Obviously there were problems to be investigated. Magendie had shown the possibility of doing long-term nutritional experiments with animals, and he continues to be credited with having provided experimental proof that animals need nitrogenous materials in their diet in order to thrive.

The Controversy over Gelatin

The next nutritional experiments to be reported were concerned with a question of considerable practical importance. Gelatin is an "animal substance," rich in nitrogen and obtainable by cooking bones in steam. It dissolves in boiling water and gives an attractive, "sticky" physical character to clear soup, or bouillon. Although flavorless itself, it can be seasoned with condiments and vegetables. In France at the end of the eighteenth century, where methods had been developed to extract gelatin from bones, it was a much cheaper source of nitrogen than was meat. It was important, therefore, to know whether its nutritional value relative to meat was in proportion to its nitrogen content. If it was, philanthropists were anxious to use it to provide economical but still nourishing soup for the poor.[37] The conclusion from the first trials using bouillon made from gelatin in a Paris hospital was that the patients seemed to do neither better nor worse than when they had received an ordinary bouillon made from boiling meat.

In 1814 there was a further report on the use of gelatin prepared from bones after their inorganic material had been extracted with hydrochloric acid. Hospital patients received a mixed bouillon, three-quarters from gel-

36. Magendie (1831), pp. 478–9.
37. Magendie (1841), pp. 237–42; Holmes (1974), pp. 6–7.

atin and one-quarter from meat. The change was well accepted, but the patients received the extra meat, spared from bouillon preparation, as a separate roast. The authors of the report considered it to be a generally accepted fact that the nutritive value of meat derived from its content of gelatin.[38] We have already seen references to starchy materials being regarded in this way because of their capacity to thicken soup and so give it a consistency closer to that of blood.

Hospitals in Paris persevered with the use of gelatin-bouillon as an economy measure, but in 1831, at the Hôtel-Dieu, a group of physicians, surgeons and pharmacists (that included Magendie) submitted a formal statement to their administration urging that the use of gelatin be discontinued. Their reasons were that substituting part of the regular bouillon with gelatin made it go putrid more quickly, that gelatin had a disagreeable taste, which reduced the patient's appetite, and that its nutritional value was unproven.[39] These complaints were made even though the gelatin was prepared each day in the hospital kitchen by extraction with steam from fresh bones.

In 1833 two French scientists published the results of a series of feeding trials with puppies.[40] They gave some of the animals bread alone, and other bread supplemented with either gelatin or meat bouillon. The puppies did better with a supplement of gelatin bouillon than with bread alone, but still failed to grow consistently and became feeble after some weeks. In one dog that died after a period of receiving the mixture, the tissues were wasted and extremely pale. The puppies that received pure meat bouillon with their bread or even a small proportion mixed into the gelatin bouillon did perfectly well. The final conclusions were, therefore, that gelatin had some positive value but was not equivalent to meat in nutritional value and had to be used in combination with other materials, as was always the case in practical human diets.

The French Academy of Sciences had, in 1831, set up a commission to investigate the subject after receiving a letter from a chemist who reported that he and a group of medical students had tried to live on bread and gelatin with little success.[41] Magendie was made a member of the commission and began a long series of feeding experiments with dogs. He did not report the results of the work until after 10 years of effort, and even then said that much more still needed to be done: "As so often in research, unexpected results had contradicted every reasonable expectation."[42]

There was no doubt that, after a few days of being offered nothing but

38. Leroux et al. (1814).
39. Petit et al. (1831).
40. Edwards and Balzac (1833).
41. Holmes (1974), p. 9.
42. Magendie (1841), p. 239.

gelatin, dogs preferred to starve. Adding fat, or flavoring the gelatin with salts and ham, would revive their interest for only a few days. However, the dogs were equally averse to eating egg white, either raw or lightly cooked, as their sole food. In separate trials with raw beef tendons or washed blood fibrin, dogs continued to eat for several weeks, but eventually lost a great deal of weight or died.

The only foods on which dogs continued to remain vigorous were raw bones (but *not* cooked bones), bouillon made from meat, meat (but not the residual fibers after extraction of lean meat in water for 24 hours) and crude wheat gluten containing some residual starch and fat.

Magendie suggested that chemists investigate what the extractable parts of meat were that combined with gelatin, albumin or fibrin to make it a complete food; it could perhaps be iron or other salts, fatty material or lactic acid. There was clearly more work to be done. Nevertheless, it seemed clear to Magendie that gelatin was not nutritious, and the reason was that it was not an "organic element" present, as such, in raw bone but a product of the reaction of bone material with water and heat.[43]

The Work of Boussingault

Yet another Frenchman to make an original contribution to our subject in the 1830s was Jean Baptiste Boussingault (1802–87). As with Magendie, it is interesting to consider whether Boussingault's originality can be linked to his unusual background and experience. He also grew up in Paris in turbulent times. His teachers regarded him as a blockhead, and he left school at the age of 14, thus losing the chance of going on to a university. But his parents, who ran a modest tobacconist's shop, bought him a four-volume textbook of chemistry, which had caught his interest, and he avidly attended free public lectures on science over the next two years. At 16 he was admitted to the School of Mines near Lyons. The school had been set up to produce knowledgeable mine managers, and the two years of training combined chemistry and mineralogy with the practice of mining. In his second year Boussingault was selected as "student-demonstrator" with permission to work in the laboratory at all times. Surprisingly, one of his interests lay in the production of glucose by the fermentation of different materials, and when still only 18 he submitted a paper on platinum, which was published in the principal French chemical journal.[44]

He was then appointed manager of a mine in Alsace, but in 1822 he accepted the offer of an academic position at the National School of Mines in Bogotá, capital of the newly independent Colombia, which was still at

43. Ibid., pp. 253–81.
44. McCosh (1984), pp. 3–12.

war with Spain. He spent the next 10 years in South America. There was little teaching to be done, but he devoted a great deal of time to exploring dangerous country with a portable laboratory, making geological reports and improving methods for extracting metals from ores. In addition, he arranged for a supply of iodine compounds to be used in the treatment of endemic goiter. Some 50 papers from this period were published in France detailing his analyses of minerals, volcanic ash and foods, as well as his meteorological observations at high altitudes.[45]

In 1835 he married a woman whom he had first met in Alsace twelve years earlier. Her father owned a farm there, and from that time on Boussingault applied his scientific mind to agricultural problems. He began a long series of quantitative experiments on the farm, with both plants and livestock. But he also made a home in Paris, where in 1837 he obtained a university teaching position and began to collaborate with J. B. Dumas on the chemical composition of organic materials.[46] He then spent half of each year in Paris and the other six months at the farm.

The Nutritional Equivalence of Plant Foods

Boussingault's first paper based on studies made at the farm was published in 1836.[47] He set forth the results of analyzing a range of animal feedstuffs for their nitrogen content. The values for those that are also used as human foods are presented in Table 2.1.

The author then developed an argument that we must follow step by step:

1. All the vegetable substances used as food have been found to contain nitrogen.
2. Magendie has shown that foods that do not contain nitrogen do not support life for more than a very limited period.
3. It is accepted, therefore, that the nutritional quality of a vegetable food resides principally in the gluten and vegetable albumin that it contains.
4. From this it follows that the nutritional value of a vegetable substance is proportional to its content of "animal substances."
5. It is true that some nitrogenous substances obtained from vegetable products are not nutritive, and are even violent poisons, but such substances are not present in appreciable quantities in plants eaten by humans or animals.
6. I am therefore expressing as "nutritionally equivalent" the relative weights of different foods that contribute the same amount of nitrogen, using average hay as the standard for animal feeds and wheat flour for human

45. Ibid., pp. 17–57; Cowgill (1964).
46. McCosh (1984), pp. 58–71.
47. Boussingault (1836).

Table 2.1. *Nitrogen content of foods and estimates of their nutritional equivalence*

Foodstuff	Nitrogen content (%)	Total nitrogenous nutrients[a] (%)	Nutritional equivalents[b]
Wheat flour	2.27	14.2	(100)
Rice	1.28	8.0	177
Corn flour	1.64	10.2	138
Potato flour	1.80	11.2	126
Barley meal	1.90	11.9	119
Dried carrots	2.39	14.9	95
Dry peas	3.40	21.2	67
Lentils	4.00	25.0	57
Haricot beans	4.08	25.5	56

[a]Estimated as nitrogen content × 6.25, i.e., assuming that they contain 16% nitrogen and are the only major nitrogen-containing materials present. This calculation was suggested by Boussingault in 1843.
[b]These are the quantities that contribute the same total nitrogen as 100 parts of wheat flour; i.e., the *lower* the number, the *higher* the nutritional value of the food.
Source: Boussingault (1836); Boussingault (1838a).

> foods. (In other words, if a food has twice the nitrogen content of wheat, only 50 grams [g] of it will be needed to replace 100 g of wheat in our daily diet.)

In the subsequent discussion, however, he went on to say that these were "theoretical equivalents," and it was important to check how far they agreed with the practical experience of stock feeders.

He qualified his conclusions slightly in a second paper, published in 1838, saying that he was doubtful whether humans could live on bread alone. This appeared to be because of the large quantity he thought would be required and the human capacity for food being so much smaller than that of herbivores. Some peoples were said to subsist on potatoes and others on rice. However, he knew that the potato-eating peasants in Alsace also consumed a good proportion of milk curds. And an authority on Indian life had written that the quantities of rice eaten there were very much greater than what a European could consume. In addition, the rice was eaten with a supplement, typically fish or legumes. When Boussingault himself had been in charge of a detachment of men (presumably in South America), he had found 1 lb. of haricot beans to be equivalent in value to about 3 lb. of rice, and that fitted with their nutritional equivalence, based on their relative nitrogen contents. Finally, he added values for manioc (i.e., cassava or tapioca) and for plantains (starchy bananas), assigning them only one-seventh the nitrogen content of wheat flour.[48] He was almost certainly the first person to draw attention to the nutrition problem associated with these Third World staples, which was to become a major concern to nutritionists in the 1950s.

What are we to think of Boussingault's idea that the nitrogen content of a food gave a measure of its nutritional value? Did his conclusions follow necessarily from the facts that he used in his argument? We have to say that they did not. Magendie's extended results (including those published in 1831) no longer supported the idea that a food could be judged by its nitrogen content, so that 60 g of beans (or of any other food containing 4% nitrogen) would have the full nutritional value of 100 g wheat (containing 2.3% nitrogen). Boussingault had no justification for narrowing down his criteria for a good food, as compared with the conclusions of Vauquelin and Fourcroy, who had written 30 years earlier that "beans are so nourishing, and capable of replacing all other food because they contain starch, an animal matter, phosphate, lime, magnesia, potash and iron. They yield at once the aliments and the materials proper to form and color the blood and to nourish the bones."[49] However, Boussingaut's hypothesis was stimulating, even if it served in the short term to get things out of perspective.

In his book *Rural Economy*, published in 1843–4 and quickly translated for British and American readers, Boussingault added some further important qualifications. First, he said: "I am far from regarding nitrogenous materials alone as sufficient for the nutrition of animals; but it is a fact that where nitrogenous materials are present at high levels in vegetables they are generally accompanied by the other organic and inorganic substances which are also needed for nutrition."[50] Then, at a later point, he wrote:

> ... in balancing diets for nitrogen we are considering only the *flesh* contained in the foods, and although it is unquestionably the principle of highest value, and the one most likely to be deficient, it is still not everything. Starch, sugar, gum and oil are indispensable as auxiliaries in the alimentation of cattle.... Potato and meadow hay brought to the same state of dryness contain the same proportions, i.e. 1.3–1.5% nitrogen, or about 8.5% of albumen and gluten, i.e. of flesh. But in the potato almost the whole of the remainder consists of starch, while in hay by far the largest proportion is woody fiber, which we presume to be inert. And this explains the higher value of the same weight of dry potato as an article of sustenance. To give our theoretical equivalents all the precision that is really desirable, it would be necessary to ascertain the quantity of organic matter that escaped digestion.[51]

48. Boussingault (1838a).
49. Vauquelin and Fourcroy (1806), p. 220.
50. Boussingault (1845), p. 387.
51. Ibid., pp. 390–1.

Table 2.2. *Changes in carbon and nitrogen contents of plants grown in sand, expressed relative to 100 g original seed dry matter*

	Clover		Wheat	
	C	N	C	N
Initial content (g)	50.8	7.2	46.6	3.5
Changes with germination	−11.4	0.0	−6.9	+0.1
Changes with growth				
2 months	+32.6	+0.7	+25.8	−0.3
3 months	+80.5	+2.6	+42.0	+0.2

Source: Boussingault (1838b).

The Fixation of Nitrogen by Plants

Boussingault was particularly interested in the source of the nitrogen in plants. In his memoirs, written late in life, he said that it was in 1832, toward the end of his time in South America, that he saw the value of Peruvian guano, "formed almost exclusively of ammonium salts," in making sterile soils fertile and that this caught his interest.[52] (Guano came from the breakdown of bird droppings where birds living on the abundant fish off the coast made their nests.)

His next farm studies addressed this question, beginning with elegant trials of the growth of seedlings in sand that had previously been heated in an oven until all the organic matter present had been burned off.[53] First, he compared clover and wheat seeds, and their change in composition as a result of germination, and then the composition of the products, some harvested after 2, and the remainder after 3, months of growth. The results are summarized in Table 2.2. Boussingault concluded that clover had the capacity to make use of atmospheric nitrogen and incorporate it into organic compounds, whereas wheat did not. In further trials of the same kind, peas also appeared capable of "fixing" nitrogen, whereas oats did not. Wheat and oats, presumably, normally obtained their nitrogen from nitrogenous compounds present in the soil. These results provided an explanation for the practical experience of farmers that cereal crops depleted the subsequent fertility of soils, whereas crops of clover or of legumes, such as beans or peas, increased it. Boussingault later confirmed, in a 5-year trial on his farm, that cereal crops yielded more nitrogen after a legume crop.[54]

52. Aulie (1970), p. 438.
53. Boussingault (1838b).
54. Aulie (1970), p. 448.

Table 2.3. *Results of Boussingault's first nitrogen balance experiments with a cow and a horse (expressed per 24 hours)*

	Cow			Horse		
	Materials	N (%)	N (g)	Materials	N (%)	N (g)
Inputs	Hay, 6.32 kg	2.4	152	Hay, 6.46 kg	1.5	97
	Potatoes, 4.17 kg dry wt	1.2	50	Oats, 1.93 kg	2.2	42
Total in			**202**			**139**
Outputs	Milk		46			—
	Urine		37			38
	Dung		92			78
Total out			**175**			**116**
Apparent balance			**+27**			**+23**

Source: Boussingault (1839a,b).

The First Nitrogen Balance Trials with Animals

Boussingault's next experiments were designed to investigate whether herbivorous animals obtained all the nitrogen they needed from their food. As discussed earlier, short-term respiration experiments had not given a consistent answer as to whether respiration involved the absorption of nitrogen, and the herbivores were thought most likely to be in need of such an ability. In 1839 Boussingault reported work on both a cow[55] and a horse.[56]

He explained in his papers that he chose levels of feed for each animal that kept it at constant weight. Under such conditions he assumed that an animal would require as much of each element every day as it lost – either by excretion or by the secretion of milk. If the output of nitrogen were more than the amount obtained from the food, it would have to be assumed that additional nitrogen was being obtained from the air; otherwise not.

The animals were kept in stalls and fed for at least a month on the same rations before the 3-day period during which their outputs were collected and analyzed. The cow yielded just over 8 liters of milk per day. The results for nitrogen are summarized in Table 2.3 We see that for each animal the food eaten apparently contained more nitrogen than left the animal each day either as a product (milk) or in its excreta. Boussingault concluded that,

55. Boussingault (1839a).
56. Boussingault (1839b).

Table 2.4. *Chemical cycles in the plant and animal kingdoms*

Materials	Action of plants	Action of animals
Neutral nitrogenous compounds, fats, sugars, starches, gum	Produce	Consume
Oxygen	Release	Absorb.
Carbonic acid, water	Take up	Produce
Heat	Absorb	Produce

Source: Dumas and Cahours (1842), p. 976.

because of the direction of the balance, it was extremely improbable that these animals assimilated nitrogen from the atmosphere since their food more than met their requirements. In fact, he believed that the results confirmed other reports that animals exhaled nitrogen incessantly. He also presented corresponding results for carbon, hydrogen and mineral balance and expressed his deep conviction that such elementary analysis could lead to rapid increases in our understanding of life itself.[57] This was the first of many thousands of feeding experiments with either animals or human subjects in which the balance of nitrogen and other elements was to be studied.

With hindsight, we can judge that the large positive balances in animals not gaining weight were an artifact. Boussingault's analyses for the feeds and the milk were in good agreement with modern values, but he kept samples of the excreta in a hot oven to evaporate them to dryness before beginning their analysis, and this must have led to a loss of some of their nitrogen in volatile ammoniacal compounds. In the following period, another method was developed in which fluids could be analyzed directly for their nitrogen content. In a rather similar experiment carried out 50 years later on a horse fed hay and oats, more than 60% of the nitrogen fed was recovered in the urine, and the animal was calculated to be in almost exact nitrogen balance.[58]

The Grand Cycles of Nature

Boussingault's quantitative work with both plants and animals confirmed the general concept that had been developing among French scientists since the 1780s. This was that the plant kingdom alone was capable of building up complex organic materials from simple starting compounds and pure

57. Boussingault (1839a), pp. 118, 125.
58. Atwater and Langworthy (1898), p. 357.

elements. The animal kingdom provided the other side of the cycle, breaking down complex materials, largely by way of oxidation, into carbonic acid, water and urea or other relatively simple nitrogenous compounds.

In 1842 Dumas and Cahours put this concept into the form of a table (see Table 2.4). It was realized that plants could use simple ammonium salts as a source of nitrogen. Why the legumes, but not cereals, could apparently also utilize atmospheric elemental nitrogen was unknown.

Summing Up

The advances in chemistry in this period made it possible to study nutrition in a quantitative way. Lavoisier has been called "the father of nutrition science." His first contribution was his recognition of the distinction between compounds that could change their character and "simple substances" (or elements), carbon, hydrogen, nitrogen, oxygen and others, which could not either appear or disappear. Therefore, once accurate methods of analysis had been developed, the metabolism of foods could be followed by means of balance studies of the elements carbon and nitrogen. Lavoisier's second basic contribution, in this context, was his recognition that combustion and respiration involved basically similar processes of oxidation, which could explain the phenomenon of "animal heat."

Magendie's and Boussingault's work indicated that animals were unable to use atmospheric nitrogen for the synthesis of the organic nitrogen compounds that, apart from water, made up the bulk of their tissues. Animals, and humans, were therefore ultimately dependent on plants for these. This led Boussingault to suggest that the nutritional value of different plant foods would prove to be proportional to their nitrogen content.

It required developments in atomic theory, to be described in the following chapter, for a further understanding of the nature of even the simple nitrogenous compounds.

3 Protein Discovered and Enthroned, 1838–1845

THE PRECEDING CHAPTER DESCRIBED the immediate implications of the advances in chemistry at the end of the eighteenth century, as well as some of the nutritional experiments that followed from them, dealing with the nitrogen balance of the animal kingdom. The work to be described in the present chapter depended, in turn, on the further chemical advances made at the beginning of the nineteenth century. The basis for these was John Dalton's *New System of Chemical Philosophy,* which contained the idea that the smallest units of particular compounds consisted typically of one or two atoms of each of the elements present.

Dalton was a largely self-taught Quaker schoolmaster in the north of England. His evidence for this "new system" was, first that every compound seemed to have exactly fixed proportions, by weight, of each element that it contained. Second, when there was more than one compound containing just two elements, then for a given weight of one element, the weights of the other element in the two compounds were in a simple proportion, such as 2:1 or 3:2. To use modern terminology, carbon monoxide has 133 g oxygen combined with 100 g carbon, while carbon dioxide has exactly twice as much oxygen, that is, 266 g per 100 g carbon. This relationship was called the law of multiple proportions.[1]

Dalton started with the assumption that, unless there was evidence to the contrary, the smallest unit of simple binary (i.e., two-element) compounds consisted of one atom of each element. Thus water consisted of one hydrogen and one oxygen atom. From the observations of others that gases combined in simple volume ratios and Avogadro's hypothesis that, in the gaseous state, each molecule occupied the same volume, modified formulas were developed. Since 2 volumes of hydrogen and only 1 of oxygen gave 2 volumes of water vapor, it seemed that the oxygen "unit" must split in two or, in other words, that the molecule consisted of two oxygen atoms (given

1. Dalton (1808); Meyer (1891), pp. 177–82.

the symbol O); expressed in the modern convention, it is O_2. Again 2 volumes of ammonia were formed from 1 volume of nitrogen (N) and 3 of hydrogen (H), so that nitrogen gas also apparently existed as a double molecule (N_2). With hindsight it would seem to follow that free hydrogen was also a doublet, to allow for a formula NH_3. However, practice was inconsistent and some chemists still thought of water as H–O rather than H_2O. This meant that, if the relative atomic weight of hydrogen was taken as 1, that of oxygen was 8, since the ratio of weights of oxygen to hydrogen in water is 8:1 and other atomic weights were also one-half of their modern values, with allowance for errors in the earlier determinations. Thus carbon (C) was 6 and nitrogen was thought at first to have either one-half (or even one-third) of the modern value of 14.[2]

The "Dark Forest" of Organic Chemistry

There was doubt, at the beginning, whether the law of fixed proportions, which applied to minerals, could be extended to the multitude of naturally occurring organic materials, that is, those containing carbon, hydrogen and oxygen or in some cases nitrogen sulfur (S), and phosphorus (P) in addition. But it began to be discovered that the simpler, crystallizable materials, particularly organic acids and sugars, gave proportional compositions that corresponded to relatively small whole numbers of atoms.[3] The Swedish chemist Berzelius played a leading role in the work on the composition of organic materials both with his own analyses and by collecting new published information in chemistry in a long series of yearbooks from 1822 to 1848.[4]

One crystallizable organic compound was urea, which had been obtained from urine by Rouelle as early as 1773 and had been found to give off half its weight as "volatile alkali" on being heated.[5] With the discovery of nitrogen, it was realized that this was the major nitrogen-containing excretion product in humans and other mammals. In 1819 Prout published an extraordinarily accurate analysis of it and concluded that it had the molecular formula $C_2H_4N_2O_2$. With modern atomic weights, this becomes CH_4N_2O.[6] In 1828 Friedrich Wöhler found that treating lead cyanate with ammonium hydroxide gave a white precipitate that also consisted of urea. This was a surprising result, because both cyanic acid and ammonia could in turn be synthesized from their elements, and urea was a product of nitrogen me-

2. Ladenburg (1900), pp. 47–66; Ihde (1964), pp. 95–123, 140–154.
3. Berzelius (1814).
4. Partington (1964), Vol. 4, pp. 147–8.
5. Rouelle (1773c).
6. Prout (1819), p. 376.

tabolism in the living body.[7] For the first time, therefore, a piece of "vital" chemistry, the synthesis of urea, could be reproduced in the laboratory.

The general view in this period was that the production of an organic molecule, whether in a plant or an animal, involved the use or even incorporation of a life force and that it would therefore be impossible to reproduce in a laboratory. This concept, which differed in detail from one school to another, had the general name of "vitalism."[8] Urea was, of course, only a final breakdown product of living tissues and Wöhler's work certainly did not constitute a refutation of vitalism, but it was still a surprise to scientists that he had been able to produce a chemical that until then had been known only as a product of the kidney.

We must introduce two more developments in organic chemistry before we can get to their nutritional implications. The first was the concept of "isomerism." This was based on the finding that two compounds with quite different properties could have the same elementary composition. Thus silver cyanate and silver fulminate both contained one atom each of silver, carbon, oxygen and nitrogen and were said to be "isomers." Again, Lavoisier had attributed the acidity of acetic acid to its high proportion of oxygen, but cellulose was found to have almost identical proportions of carbon, hydrogen and oxygen though with no trace of acidity. The only explanation for such differences was that the atoms were combined in different ways in the different isomers and that knowing the overall proportions of elements was not sufficient to predict the properties of compounds.[9]

The second discovery was that in organic chemistry some atomic groupings, or "radicals," seemed to persist and to behave almost as if they were elements in inorganic chemistry. One was the gas cyanogen (formed from carbon and nitrogen), which formed an acid and salts and reacted with organic compounds in a manner similar to chlorine.[10] Another was the benzoyl radical. Distilling bitter almonds with water yielded an oil that on exposure to air formed crystals of benzoic acid, which had already been obtained from resins of the benzoin laurel tree. The oil formed more compounds with chlorine or bromine and also a cyanide derivative. But benzoic acid was re-formed from all these on reaction with water. It appeared that the "benzoyl" group (with the formula C_7H_5O) was a stable radical, though if the compounds were ignited in oxygen they were fully oxidized to carbon dioxide and water. This work, by Wöhler and Liebig, was hailed by Berzelius as beginning a new day in vegetable chemistry and "opening a way into a dark forest."[11]

7. Wöhler (1828); Ihde (1964), pp. 164–5.
8. McKie (1944); Toulmin and Goodfield (1962), pp. 322–6; Teich (1965), 44–5.
9. Partington (1964), Vol. 4, pp. 256–7.
10. Ibid., pp. 253–4.
11. Vickery and Osborne (1928), p. 394; Holmes (1973), p. 333.

The Protein Radical

This preliminary review brings us up to the year 1838, where the preceding chapter broke off, and sets the scene for a famous paper published by Gerrit Mulder, a Dutch physician who had taught himself chemical analysis.[12] He reported analyses for egg albumin, serum albumin, fibrin and wheat gluten and concluded that they contained an identical common radical. The molecules of the individual products differed only because they were combined with different numbers of sulfur atoms and, in some cases, phosphorus atoms. He said that these extraneous atoms could be removed from the naturally occurring nitrogenous materials by treatment with potassium hydroxide. The radical could be precipitated by adding acetic acid. He assigned it a formula of $C_{40}H_{62}N_{10}O_{12}$ on the basis of the reaction product obtained from treating egg albumin with strong, cold sulfuric acid. He assumed that one molecule of the acid had combined with one molecule of the radical, later given the symbol Pr. However, in both lead and silver salts of albumin, the proportion of the metals was much lower, so that one atom was apparently combined with 10 molecules.[13] When purified serum or egg albumin or fibrin was analyzed, each was also found to contain sulfur and phosphorus at such low levels that again it appeared that the formulas were as follows:

Serum fibrin and egg albumin: $Pr_{10} \cdot SP$
Serum albumin: $Pr_{10} \cdot S_2P$

Mulder sent a summary of his findings to Berzelius, who gave an enthusiastic reply and suggested a name for the new radical:

The name *protein* that I am proposing for the organic oxide of fibrin and albumin I chose to derive from the Greek word πρωτεοδ (proteios), because it appears to be the fundamental or primary substance of animal nutrition which plants prepare for herbivores, and who in turn supply it to the carnivores. If one were to choose a name based on the Greek word for fiber it would be less appropriate since the organic oxide is also the base of albumin [in addition to fibrin] and probably of the coloring matter [hemoglobin] as well as of still other materials.[14]

The ancient Greeks had a god, Proteus, who had the power to change himself into different shapes. Mulder followed Berzelius's suggestions for the name of his new radical, and expressed his conclusions as follows:

It seems certain that animals obtain their basic materials from the plant kingdom. It is possible that plant albumins have proportions of sulfur and phosphorus that differ from those in animal albumins, but the ternary [four-element] organic body is "*protein*" in each case.... It is not yet known whether starch and other substances

12. Brouwer (1952), p. 4; Glas (1975), pp. 291–9.
13. Mulder (1839), p. 138; Snelders (1982), pp. 200–1.
14. Vickery (1950), p. 389.

which have been found to be of nutritive value, can also be converted into protein in the animal body.[15]

Joining the Bandwagon

The idea was well received by other chemists. In 1841, in the *Annalen der Chemie und Pharmacie,* of which he was the editor, Justus Liebig compared the analyses of nitrogenous materials extracted from plants with analyses reported from Mulder's laboratory for serum albumin, casein and fibrin. He concluded that they all contained a ratio of exactly 8 equivalents of carbon to 1 of nitrogen.[16] (Because of the continuing controversy over atomic weights, Liebig had reverted at this time to "equivalent weights" and was using 6 for carbon and 14 for nitrogen.) Therefore, with modern atomic weights of 12 and 14, respectively, this corresponded to a ratio of 4 atoms of carbon to 1 of nitrogen, which was in agreement with Mulder. He also said that when "the casein of sweet almonds or legume grains" was suspended in an alkaline solution (potassium hydroxide) with gentle heat and then acidified, there was an abundant production of hydrogen sulfide and a precipitate of protein.[17]

Liebig recorded in his paper some general thoughts: "The similarity of the vegetable nitrogenous compounds to the essential parts of the blood is not confined to their chemical composition; their chemical reactions are also the same." Later he added that the inorganic elements adhering to them, magnesium, calcium, iron, sulfur and others, were also the same. "A carnivore eats itself, from a chemical point of view, because its nourishment is identical with the component parts of its body; but a herbivore does the same, because its foods are also identical with its flesh or its blood."[18] Elsewhere he wrote: "It may be laid down as a law, founded on experience, that vegetables produce, in their organism, compounds of protein; and that out of these compounds of protein the various tissues and parts of the animal body are developed by the vital force, with the aid of the oxygen of the atmosphere and of the elements of water."[19]

In 1842 Dumas and Cahours, working in Paris, reported that they had followed Mulder's procedure for the preparation of protein from both casein and serum albumin. In each case the products had a composition that corresponded best (after conversion to modern atomic weights) to the formula $C_{48}H_{74}N_{12}O_{15}$,[20] which again has a ratio of exactly 4 atoms of carbon

15. Mulder (1839), p. 140.
16. Liebig (1841), pp. 145–9.
17. Ibid., p. 152.
18. Ibid., pp. 151, 154.
19. Liebig (1842a), p. 102.
20. Dumas and Cahours (1842), pp. 985–6.

to 1 of nitrogen. However, the French workers differed from Liebig in finding that the nitrogenous material, "legumin," that could be extracted from peas and beans, and that Liebig called "vegetable casein," contained only 3.25 atoms of carbon per atom of nitrogen. They added that, although legumin therefore had a chemical composition closer to that of gelatin, in view of the proven nutritional value of legumes (in contrast to the apparent worthlessness of gelatin), it must really be a compound of albumin with some other product in fixed proportions.[21]

Dumas and Cahours also discussed the metabolism of protein in the body.[22] An adult man consumed each day 100 to 125 g "neutral nitrogenous material," containing 16 to 21 g nitrogen. Almost the same quantity of nitrogen was recovered in the urine each day in the form of urea, which could be regarded as an oxidation product of albumin or casein. Ignoring intermediate steps, the overall reaction could be represented (after converting to modern atomic weights) as

$$C_{48}H_{74}N_{12}O_{15} + 100\text{ "O"} \rightarrow 6\ CH_4N_2O\ \text{(urea)} + 42\ CO_2\ \text{(carbonic acid)} + 25\ H_2O\ \text{(water)}$$

From an average daily intake of protein, 50 g carbon and 6 g hydrogen were not recovered in urea, but were combusted to carbonic acid and water, respectively. Since 1 g carbon had been found to give 7.3 kilocalories (kcal) on combustion and hydrogen gave 35 kcal/g, the total daily heat from the combustion amounted to 575 kcal. However, it was known that a man produced 2,500 to 3,000 kcal of heat each day and that this sufficed to maintain his body temperature. With only 575 kcal, he would die of cold. Therefore, he needed to burn another 240 g fat per day, or its caloric equivalent (2,200 kcal) in starch or sugars. This explained a man's nutritional needs for non-nitrogenous organic materials in addition to the nitrogenous ones. (1 kcal is the amount of heat needed to raise the temperature of 1 kilogram [kg] water by 1°C.)

In the same year (1842), Liebig set out a series of 20 chemical equations to explain the production of different materials in the body.[23] Thus he attempted to demonstrate that 2 molecules of protein could be broken down to yield 1 of fat and 6 of urea. He also wrote that 1 molecule of gelatin appeared to be made up of 2 molecules of protein, 3 of ammonia, 1 of water and 1 of oxygen and drew the conclusion that "by means of these formulae we can trace the production of the different compounds from the constituents of blood."[24]

Berzelius was alarmed at these "speculations to describe the chemical

21. Ibid., pp. 992–4.
22. Ibid., pp. 998–9.
23. Liebig (1842a), p. 147.
24. Ibid., pp. 121–2.

phenomena which occur in living bodies.... readers will be misled into regarding probabilities as truths, and it will be the more difficult later on to eradicate errors."[25] But despite this criticism of speculative chemical physiology, it was generally accepted by 1842 that the different nitrogenous compounds in animal and plant tissues for which there had been no unifying name were all based on a common large unit, or radical, called "protein."[26]

The Position of Justus Liebig

In this period Liebig was such an influential and controversial figure in the fields of both animal and plant nutrition that it seems useful to review how this came about. He was born in 1803 in Darmstadt (some 20 miles south of Frankfurt, Germany). He was not very successful at school, where the principal told him that he was "the plague of his teachers and the sorrow of his parents." However, he became interested in chemistry while helping his father, who was a wholesaler, and to a small extent a manufacturer, of dyestuffs. At 16, he managed to gain enrollment at the University of Bonn, where there was a professor of chemistry. The chemistry taught there was very theoretical, but after graduation at 19 Liebig obtained a grant from the Grand Duke of his area to study chemistry abroad. The choices open to him were Berzelius's modest laboratory in Sweden and the University of Paris with a variety of laboratories in its neighborhood. Liebig chose the latter and greatly enjoyed the French tradition of lecture demonstrations. He was lucky enough to catch the attention of Gay-Lussac and to become his personal assistant in the latter's research laboratory for nearly two years.[27] Gay-Lussac was experienced in the analysis of organic compounds and interested in their fine structure as a result of observations of isomerism. He was more concerned with the discovery of general laws than with the proliferation of individual observations.[28]

At the early age of 21 Liebig was appointed assistant professor in the small University of Giessen, some 30 miles north of Frankfurt. Here he set up a laboratory, at first in an unused barracks, where his students could work on individual research projects under his supervision (Figure 3.1). This was the first facility of its kind anywhere, and Liebig soon attracted students from many countries. Here he generated a previously unheard of rate of experimentation, principally on the preparation and characterization of organic compounds. Liebig was also extraordinarily active in teaching and writing as well as in editing his own *Annalen der Chemie*.[29]

25. Berzelius (1842), pp. 535–6.
26. Fruton (1972), pp. 97–8.
27. Shenstone (1901), pp. 11–19; Partington (1964), Vol. 4, pp. 294–5; Munday (1990).
28. Crosland (1978), pp. 87–136.
29. Browne (1942); Partington (1964), Vol. 4, pp. 296–7.

Figure 3.1. "Liebig In His Laboratory at Giessen" as portrayed on an advertising card issued with boxes of Liebig's Meat Extract. (Dr. P. A. Munday)

In 1837 Liebig was invited by a group of his old pupils to England, and gave a lecture to the British Association for the Advancement of Science. In his talk he stressed the practical applications of organic chemistry, particularly in connection with the agricultural advances in Britain and the work in progress there, in both animal and plant physiology. The association expressed great interest in his ideas and asked him to prepare a report.[30] This has frequently been taken as the stimulus that completely changed Liebig's interests and caused him to write *Organic Chemistry in Its Relation to Agriculture and Physiology*, which appeared in 1840. However, a modern scholar has argued that Liebig had always been interested in the practical applications of chemistry and that because of his fascination with the metamorphosis of materials, combined with the new information linking the components of plants and animals, it was a natural progression to consider the synthetic powers of plants.[31]

Some, like Berzelius, thought that plants had only a limited ability to produce living material out of purely inorganic substances and that it was therefore the soil's complex organic materials – that is, the humus fraction

30. Rossiter (1975), pp. 27–8; Brock and Stark (1990), pp. 134–5.
31. Munday (1991).

from decayed plant residues – that was important for new crop growth. Liebig did not see any fundamental distinction between inorganic and organic chemistry, nor did he feel that a plant's power of synthesis was limited in this way. He believed that what plants needed from the soil was water, minerals and ammonia – the last being supplied by rainwater, in which it naturally dissolved. Fertilizers would increase crop growth if they supplied minerals in short supply in soils. These ideas were set out in his 1840 volume. The direct, authoritative tone of the book helped to make it a "best-seller." It was translated into many languages and was particularly influential in the United States.[32]

Liebig's ideas led him to take a financial interest in the British production of "patent fertilizers" carrying his name. They were apparently marketed without prior testing because of his certainty that they would be effective. In fact, they were a fiasco, because they were too insoluble to have any worthwhile effect in the short term.[33] The subject of fertilizers is not directly related to this book, but it shows the range of Liebig's activities and illustrates his willingness both to plunge fearlessly into a new field and to set out dogmatic opinions that were not always correct.

PROTEIN: THE ONLY TRUE NUTRIENT

After completing his book on plant nutrition in 1840, Liebig turned his energies to the subject of animal (and human) nutrition, although he had never carried out a feeding trial or any kind of metabolism experiment with animals, and only one rough balance trial on a group of soldiers. He was spurred on, it seems, by his rivalry with Dumas and Boussingault in Paris and his fear that they would publish and gain priority for ideas that he was already introducing in his lectures. The first edition of *Animal Chemistry* was published in 1842 and was immediately translated into English. It has already been the subject of a thorough study.[34] I will consider it only in relation to its direct implications for the science of nutrition. The main points of Liebig's argument can be compressed into the following statements (using my words rather than his):

1. The bodies of stall-fed domestic animals may contain fat, but it is not an essential component since animals in their natural, wild state do not accumulate fat. In nature it is found only as a relatively small component of an animal's brain and nerves.[35]

2. The dry matter of the lean functional tissues contains approximately

32. Liebig (1840).
33. Rossiter (1975), p. 44; Finlay (1991), pp. 157–8.
34. Holmes (1964),
35. Liebig (1842a), pp. 80, 85–6.

17% nitrogen. This is also the nitrogen content of the two principal components of the blood plasma (i.e., the blood without its globules, which are assumed to be oxygen carriers and to play no part in nutrition). These components are albumin and fibrin, which differ in some properties and therefore have a different arrangement of atoms. However, they have the same chemical composition, and it has been demonstrated that they can be converted to one another. The conversion of the constituents of blood into muscle fiber (and vice versa) involves, again, only a change of form. Muscle fiber, and in fact all organized tissues, must of course be formed from the blood that bathes them.[36]

3. The only true nutrients are therefore those that can be converted to blood. In carnivores the process of nutrition is easily understood. The nitrogenous tissues of the meat consumed are solubilized by the digestive juices, and if necessary by a type of fermentation that gives the rearrangements needed for their conversion to soluble albumins. These are then absorbed directly and become part of the blood.[37]

4. The food of grain- and grass-eating animals is apparently very different, but it too contains vegetable fibrin, albumin and casein, with a composition identical to the animal components, and they become integrated into the animals' tissues exactly as in the carnivores. The other nitrogenous constituents of plant foods are rejected by the animals.[38]

5. The functions of the non-nitrogenous components that exist in larger quantities in these animals' foods will be discussed later, but it is certain that they cannot be "animalized" by reaction with atmospheric nitrogen.[39]

6. The turnover of proteinaceous foods by adult animals, as shown by the continued excretion of urea even when none of these materials is consumed, is explained by muscles consuming themselves when they exert their muscular force. This force is released by the molecule breaking into two fragments. One, rich in nitrogen, is broken down to form urea, together with carbonic acid and water. The urea is excreted in the urine, and the amount of mechanical force exerted by a man or animal in a particular period is proportional to the quantity of urea excreted. It cannot be supposed that the nitrogen of food can pass into the urine as urea without having previously become part of an organized tissue.[40]

7. The second type of fragment formed during muscular exertion is carbon-rich; it is also released into the blood and picked up by the liver, converted to choleic acid and secreted into the bile. When the bile, in turn,

36. Ibid., pp. 39–42, 220.
37. Ibid., pp. 43–4.
38. Ibid., pp. 44–7.
39. Ibid., pp. 71–2.
40. Ibid., pp. 231–3.

is secreted into the gut, the choleic acid is largely reabsorbed and oxidized to carbonic acid and water.[41]

8. The breakdown of muscle that occurs during the day is compensated for by re-formation of tissues during sleep, and the "force" or vitality of the muscles is regained. For an active adult, 7 hours of sleep are required. An old man, who is necessarily less active, requires only 3½ hours. In each case "waste is in equilibrium with supply." However, in the infant who sleeps 20 hours and is awake for only 4, there is an excess of supply, and this explains the child's ability to gain weight and grow.[42]

9. Since only those substances that are capable of conversion to blood can properly be called nutritious, or considered to be food, the protein elements of foods are the only true nutrients, that is, the only ones capable of forming or replacing active tissue. The gelatin of the skin and connective tissue is not itself a form of albumin. It can be formed from it, but the reverse process does not take place. Gelatin in the diet can thus spare the small proportion of dietary protein otherwise needed for the production of skin and connective tissue, but it cannot meet the larger need for replacement of muscle fiber.[43]

10. The non-nitrogenous components of the diet also have a role, which is to support the process of respiration. They fall into two main classes. The first includes starch, gum and the sugars. Their chemical composition can, in each case, be represented as so much carbon and so much water. Also, on digestion or fermentation, starch can be converted to sugar. The second class of materials comprises the fats, and these can also be formed in the body from sugar when the latter is being ingested at a faster rate than is needed to react with oxygen, that is, when there is a relative shortage of oxygen.[44]

11. The oxygen of the air serves two vital functions in animals, but it can also be damaging. Its first function is to take part in the digestive reactions in the stomach, and it is carried there by saliva, whose mucus has the special property of forming bubbles that trap air. The surplus nitrogen in these bubbles is reexcreted through the skin. The second function of oxygen is to take part in the reaction by which muscle fibrin decomposes, as described earlier, with the release of mechanical force. The danger is that too much oxygen may enter the circulation and cause excessive oxidation and loss of protein tissue. This can be prevented only by non-nitrogenous foods reacting preferentially with the excess oxygen. The secondary effect of both types of oxidation is the production of necessary animal heat.[45]

41. Ibid., p. 58.
42. Ibid., pp. 236–7.
43. Ibid., pp. 38, 122–4.
44. Ibid., pp. 68–71, 78–81, 85–6, 90.
45. Ibid., pp. 2, 108, 209, 232.

12. The special role of body fat in protecting the essential, nitrogenous body tissues is clearly shown by the course of events during starvation when foods are no longer present to protect the system from the oxygen of the air. First, there is gradual oxidation of body fat. When this is exhausted, the muscles become shrunken as the albuminous structure is gradually oxidized. Finally, the brain and nerves are attacked and death ensues.[46]

13. The rate at which oxygen enters the body, and thus the rate at which sugars and fats have to be used up to protect the tissues, depends on the rate at which it enters the lungs. At higher atmospheric temperatures the air is thinner and, where the climate dilutes the air with moisture, there is in each case less oxygen in a single respiration. This explains why an Englishman transferred to the hot, moist climate of Jamaica experiences a loss of appetite; the carbon and hydrogen of his food are oxidized at a lower rate. Fortunately, the foods most abundant in the tropics are watery fruits. Conversely, someone living in the cold of the Arctic and inspiring more concentrated air has a much greater appetite, and nature has arranged that the foods there are concentrated and fatty. It is also part of nature's arrangement that the higher rate of metabolism produces proportionately more heat, which is of course required to maintain body temperature in that climate.[47]

14. Someone who feels cold is induced to engage in physical activity. This stimulates respiration, part of which is needed for the breakdown of muscle fibers, but it admits more oxygen, which also results in more combustion of the non-nitrogenous protectors and, thus, in more heat production. As Lavoisier showed, physical activity results in greatly increased carbonic acid production.[48]

15. For the carnivore who eats only lean meat, the oxygen can react only with organized tissues. Thus lions and tigers housed in cages in a zoo are in incessant motion so as to furnish the "material necessary for respiration." Similarly, a savage living only on meat is forced to make the most laborious exertions.[49]

Obviously, in a book of more than 300 pages Liebig said a great deal more, but this summary presents the principles of human and animal nutrition as he saw them. Modern readers probably see the scheme as so "upside-down" as to be painful. Some of its implications are seen in Lyon Playfair's paper delivered to the Royal Agricultural Society in Britain, immediately after the publication of *Animal Chemistry*. Playfair had worked, as a student, in Liebig's laboratory and had translated his earlier volume

46. Ibid., p. 25.
47. Ibid., pp. 15–17, 21–3.
48. Ibid., pp. 20, 85–6.
49. Ibid., p. 75.

into English. Now he said that Liebig's contribution had important practical applications for animal breeders. Previously, animals had been selected for well-developed lung capacity. However, now that it was understood that allowing a larger quantity of oxygen into the body only resulted in a greater oxidation and loss of sugars and fat, it was obviously desirable to select narrow-chested animals for breeding, because these would take in less oxygen with each breath, and thus would burn less carbon and hydrogen, and so leave more available for fattening.[50] The conclusion was logical enough, given Liebig's premise, but what an extraordinary one it was, with its implication that we would all be better off half-choked, so that less oxygen reached our lungs.

The Reception of *Animal Chemistry*

The book greatly impressed many of the general reviewers in medical journals. They commonly questioned particular statements, but the work as a whole was thought to represent a step forward into a new era.

Liebig had derided physiologists as being in no position even to comment until they had also made chemists of themselves. One of them wrote a whole book in reply: *The Inter-relationships between Physiology and Chemistry as Illuminated by a Critique of Liebig's "Animal Chemistry."*[51] The author contrasted the typically meat-rich diet of the sedentary, well-off businessman with the cheaper, high-carbohydrate diet of the day laborer who earned his smaller income by the sweat of his brow. How could this be reconciled with Liebig's dogma of albuminoids being the only source of muscular energy and of physical activity being the only way in which they could be broken down and eliminated? An alternative hypothesis, which fitted the facts better, was that the physically active person on a high-carbohydrate diet was able to keep going because, as muscle tissue broke down, the nitrogen-rich fragments recombined with carbohydrates to provide replacement muscle material, so that there was no longer any clear differentiation of function between albuminoids and respiratory foods.[52] It had already been shown that, as seemed inevitable, very large amounts of urea could be excreted by someone on a diet rich in albuminoids without any increase in muscular activity.[53]

J. R. Mayer, now remembered mainly as one of the principal independent discoverers of the conservation of energy, also criticized Liebig for his inconsistencies and contradictions concerning the relationships between mus-

50. Playfair (1843), pp. 257–9.
51. Kohlrausch (1844).
52. Ibid., pp. 51–2.
53. Lehmann (1842), pp. 273–4.

cular work and heat production. The oxidation of carbon compounds must be the source of chemical energy for muscular contractions, and this appeared partly in the form of mechanical energy and partly as heat. Using a rough value for the mechanical equivalent of heat, as well as estimates of the amount of mechanical work done by laborers and the extra food they needed to eat as compared with men at rest, he calculated the mechanical efficiency of the muscular "engine" as about 33%.[54] However, Mayer was still considered something of a crank at this time, and his fundamental contribution was not to be appreciated for another 10 years.

The French Challenge to Liebig

Dumas and Boussingault agreed with Liebig as to the primary importance of the nitrogen-containing components of food, but were angry with him for having claimed priority for discovering the identity of plant and animal materials – the albumins, caseins and so on. But, ironically, the claim in *Animal Chemistry* that they chose to challenge was one of the few points in Liebig's scheme that a modern reader would consider to be correct.

The French workers' summary of the relative roles of the plant and animal kingdom in nature (Table 2.4) included the generalization that plants were capable of "reduction" reactions – that is, those adding hydrogen or yielding free oxygen – but that animals were capable only of oxidation, which used oxygen and resulted in the breaking down of organic compounds. Now the conversion of starch or sugar to fat came into the category of "reduction," as Liebig himself had said.[55] Using one of his reaction schemes (converted to modern atomic weights), it could be expressed as

$$\underset{\text{Starch unit}}{(C_{12}H_{20}O_{10})} \rightarrow \underset{\text{"Fat"}}{C_{11}H_{20}O} + CO_2 + 7\text{ "O"}$$

Dumas and Boussingault believed that Liebig had ignored a general law that they had discovered, and set out to prove him wrong. Some of their own colleagues in the Academy of Sciences warned them against being so dogmatic about the limited powers of synthesis by animals. Magendie stressed that "nutrition remains one of the most obscure questions in science," and another member wrote that it would seem to follow from their law that snakes could secrete venom only when they had obtained it in their food.[56]

Boussingault and others in France were stimulated to do very careful balance trials with successive carcass analyses, but finally had to admit, in

54. Mayer (1845), pp. 52, 57–8.
55. Liebig (1842a), pp. 86–7.
56. Holmes (1974), pp. 48–117.

1845, that fat could be synthesized from starch or sugars. As another critic was to put it some 20 years later, "This decision in his [Liebig's] favour contributed greatly to the extension of his reputation."[57] The French scientists had lost the debate that had been waged on ground of their own choosing. Dumas continued to have a distinguished career but withdrew from the field of nutrition, and Boussingault concentrated his subsequent work on crop husbandry and the problems of nitrogen fixation in the soil.[58]

Summing Up

By this period it had begun to be shown that even organic compounds obeyed the "law of fixed proportions," and atomic proportions had been worked out for urea, benzoic acid and other compounds. Then in 1838 Gerrit Mulder claimed to have discovered the existence of a protein radical with the formula $C_{40}H_{62}N_{10}O_{12}$ and that this combined with sulfur and phosphorus to form albumin, fibrin and so on. Justus Leibig adopted this concept and went further, concluding that the protein radical, a chemical product of the plant kingdom, was the only true nutrient. It was therefore the essential ingredient for both body building and physical activity. This was an elegant concept and apparently was firmly based on recent developments in both organic chemistry as well as a multitude of observations from animal experiments. The broad conclusions were to survive in people's minds for many generations, but the scientific bases for them were to disintegrate quite quickly, as we shall see in the next chapter.

57. Voit (1870), p. 371.
58. Kapoor (1971); McCosh (1984), pp. 248–59.

4 Things Fall Apart, 1846–1875

THERE WAS LITTLE CRITICISM from 1842 to 1846 of Liebig's scheme of *Animal Chemistry* outside a few German university circles. It appeared to be a giant intellectual synthesis. And coupled with Mulder's discovery of "protein" as the common repeating unit in the molecules of the albuminoids, as well as related nitrogenous compounds, it seemed to introduce an attractive harmony into the principles of nutrition and to have explained its basic principles. However, this scheme was not to last.

THE DISAPPEARANCE OF "PROTEIN"

As a result of further work in his laboratory, Liebig published a note in his *Annalen* early in 1846 saying that he and his colleagues had been unable to prepare sulfur-free protein from any of a variety of starting materials; he ended by asking Mulder to give precise details of how he had accomplished this and what tests he had conducted to check that his products were truly free of sulfur.[1]

A detailed paper by Laskowski, a Polish visiting scientist in Liebig's laboratory who had done the bulk of the work, appeared later that year. He had started with egg albumin and treated it with potassium hydroxide in solutions of different strengths and at different temperatures. This dissolved the albumin. Following Mulder's instructions he then acidified the solution, which usually resulted in a precipitation of an albumin-like material. However, these precipitates, when tested by drying and heating with solid potassium hydroxide, and then re-acidifying and heating again, all gave off hydrogen sulfide (the gas with the characteristic smell of bad eggs); and paper, moistened with lead acetate solution and held over the mixes, turned through the formation of lead sulfide. If the original alkaline albumin solution was heated at its boiling point for a period before acidification, the

1. Liebig (1846a).

albumin did lose much more sulfur as hydrogen sulfide, but then no precipitate was formed – the albumin having presumably undergone a greater degree of decomposition.[2]

Laskowski went on to describe other experiments with casein, and also presented analyses of egg albumin blood fibrin and other materials that did not agree with Mulder's models of these being a combination of 10 to 15 molecules of protein with 1 to 2 atoms each of sulfur and phosphorus.[3] His final conclusion was that there was no basis left for believing in the existence of Mulder's fundamental substance, "protein."[4]

Mulder was enraged by Liebig's condescending attitude and wrote urging him to apologize publicly in the *Annalen*.[5] When he received no reply, he hurriedly published a booklet denouncing Liebig both for his science and his character, defending the existence of "protein" and proposing a modified method for its preparation.[6] Mulder's reply attracts some sympathy from the reader. Liebig apparently considered himself to be in a superior position that allowed him to sit in moral judgment over others. He had already called Dumas a plagiarist who had been wasting his time in worthless speculations and had described Laurent and Gerhardt, two scientists still honored for their contributions to organic chemistry, as conceited, self-complacent cocks strutting about on a dung hill.

Mulder added that his original paper, written in Dutch, gave full experimental details and that it was only an abbreviated translation, appearing in a German journal, that omitted them. He also pointed out that Liebig had himself referred to "protein" having been obtained in his laboratory and, in a rebuttal to Dumas, had claimed a formula of $C_{48}H_{74}N_{12}O_{15}$ for it, based on its supposed decomposition to choleic acid and uric acid.[7] However, Mulder's revised claims still could not be confirmed by other chemists. His ideas were therefore discarded, and English-peaking authors went back to the term "albuminoids" when referring to the group of nitrogenous materials as a whole.

Liebig's Revised Views

The first edition of *Animal Chemistry* was, of course, prepared when Liebig was a believer in Mulder's "protein radical." In the 1846 edition, after his rejection of the concept, he omitted all mention of it and simply restated: "It was established as a universal fact that the sulfurized and nitrogenized

2. Laskowski (1846), pp. 152–8.
3. Ibid., pp. 133–9.
4. Ibid., p. 165.
5. Snelders (1982), p. 212.
6. Mulder (1846).
7. Scherer (1841), p. 44; Liebig (1842b), p. 355.

constituents of vegetable foods have a composition identical with that of the constituents of the blood."[8] He seemed not to consider the growing knowledge of digestion as worthy of mention. Presumably he was still of his former opinion that "in the digestive process the food only undergoes a change in its state of cohesion without any other change of properties."[9] We will return to the subject in Chapter 7.

In 1847 Liebig took another sharp turn. In his *Researches on the Chemistry of Food,* he began by lamenting how few securely based facts there really were in animal chemistry. "This was because the subject had been in the hands of adventurers.... The protein theory had no foundation and was never regarded by those intimately acquainted with its chemical groundwork as an expression of the knowledge of a given period."[10] This was going too far. He had chosen, consciously or unconsciously, to forget his own enthusiastic use of the concept in his earlier writings. And, in fact, in the very next section he said that it was under the influence of Mulder's authority that he had come to believe that fibrin, casein and albumin all had exactly the same chemical composition and to ignore results, from his own and many other laboratories, that indicated the opposite.[11]

So he now withdrew what he had said only the preceding year to be "established as a universal fact" and wrote that it was, after all, not necessary for all the components of blood to have been already present in food. Even if food contained three different compounds – one containing nitrogen, another sulfur, and a third additional carbon, they could all react together finally to form blood if the appropriate attractive forces preponderated.[12] As has been pointed out already, this nullified all his former arguments that carbohydrates and fats were purely respiratory and had no role in the production of tissues.[13] This may have come home to Liebig also, because the promised Part II of his 1846 edition of *Animal Chemistry* never, in fact, appeared.

For the next 20 years Liebig kept a relatively low profile on the subject of nutrition, except for his proprietary "extract of meat" (Figure 4.1) and "malt food for babies," which we will discuss later in the chapter. He also gave up supervising an experimental laboratory with his move to Munich in 1852. However, he never publicly withdrew his view that non-nitrogenous foods were purely "respiratory" and that only albuminoids were truly nutritional.

8. Liebig (1846b), p. 135.
9. Liebig (1842a), p. 115.
10. Liebig (1847), pp. 1, 2, 19.
11. Ibid., p. 21.
12. Ibid., pp. 19–21.
13. Holmes (1964), p. lxxxix.

Figure 4.1. "Liebig Is Summoned to Meet King Max" as portrayed on an advertising card issued with boxes of Liebig's Meat Extract. (Dr. P. A. Munday)

Critical Comments

Liebig's writings were widely circulated and rapidly translated into other languages. His critics were not as widely read, but they could be biting, particularly in the light of the changes in his ideas between successive publications and his withdrawal of many of the hypothetical chemical equations found in the first edition.

The following are typical comments from an English review:

A bold speculator will never be in want of earnest admirers and enthusiastic followers ... but his statements appear to us to require impartial examination at the hands of men who have no hypotheses to support. To take one instance, he refers to the consumption by people living in the cold of the far North of much blubber and other kinds of highly concentrated fatty food. Liebig says this is absolutely necessary to them to prevent the combustion of their bodies by the condensed state of the oxygen in the cold air which they breathe. But the reindeer living in the same area defy the Arctic oxygen while living on dry, farinaceous moss and other plant foods. There is an entire want of proof that man's food is necessarily regulated by the variable proportion of oxygen contained in a given bulk of the atmosphere.[14]

14. Anonymous (1846).

In the subsequent decade, interest in metabolism and nutritional questions was developing in the United States. In 1856 a 21-year-old medical student, John C. Draper, published the short thesis that he had written for his medical degree at the New York University Medical College. Its title was "Is Muscular Motion the Cause of the Production of Urea?" He said that he had kept his diet constant, that on a day of rest he excreted 26.4 g urea and that on two active days (one of them including a 13-mile walk) the excretion was not significantly different.[15] His father, John W. Draper, was president of the college and had developed a new method of quantitative analysis for urea a few years earlier.

The paper was certainly referred to by European workers. However, neither of the Drapers extended his work, and another paper published in the United States in the preceding year had given contradictory results. W. A. Hammond, who was to be a controversial surgeon-general for the northern armies at the beginning of the Civil War, was interested in Liebig's belief that the first product from the breakdown of protein was uric acid and that urea was produced from it in its turn. He reported that on a day of rest he excreted 1.6 g uric acid and 31.6 g urea. On another day of strenuous exercise he found only one-third as much uric acid in his urine, but nearly twice as much urea.[16] He too reported no further experiments. The United States was not yet an important source of scientific or medical advances, and since the two papers appearing from this country largely neutralized each other, it is not surprising that little notice was taken of them.

The British Approach to Nutritional Questions

We have already referred to the work of one English scientist active in physiological chemistry in the first half of the nineteenth century. This was William Prout, who lived from 1785 to 1850 and received his training in medicine at Edinburgh University. He was then attracted to organic chemistry and to the atomic theory. In particular, he became convinced that there must be a special significance to the fact that so many atomic weights (e.g., for carbon, oxygen and nitrogen) were whole-number multiples of the atomic weight of hydrogen. He also demonstrated that the acid secreted into the stomach was hydrochloric acid and is credited with having been the first to divide the principal organic constituents of foods into the three categories of fats, carbohydrates and albuminoids. He was a religious man and argued that God would not have included members of all three groups in milk if they did not each have an essential nutritive function.[17]

15. Draper (1856).
16. Hammond (1855).
17. Brock (1975).

Prout's approach was very much that of the basic scientist-philosopher, but the typical approach to nutritional problems in Britain at this period was more applied. First there was a continuing interest in England in applying scientific knowledge to the improvement of farm animal production, through both selective breeding and efficient husbandry. Lawes and Gilbert, who had already publicly disagreed with Liebig over the value of nitrogenous fertilizers, carried out detailed animal feeding studies at the Rothamsted Experiment Station in the 1850s. From their work on the growth of sheep and pigs on different diets, they concluded that the importance attached to the nitrogenous fraction of feedstuffs by both Boussingault and Liebig had been greatly exaggerated. They commented further that they had been "unable to discern a sufficient basis of facts" for the evaluation of human diets in terms of their nitrogen content. It appeared to them that people doing hard physical work instinctively increased their intake of fat and other non-nitrogenous materials.[18] In addition, they pointed out that their detailed studies of the composition of carcasses of animals killed in the "fat" condition desired by the market showed that they actually had a lower ratio of nitrogen to carbon than was present in bread. Fat meat could not therefore be considered to be primarily a source of albuminoids. They suggested that the proven value of meat as a nutritional supplement to bread must therefore have some other, still unrecognized basis.[19]

A second practical problem in Britain was the feeding of people who were the responsibility of the state. Members of Parliament, and others interested in social problems, were concerned that prisoners and paupers in institutions should receive food adequate to support health but not as "luxurious" as the food that even the poorest independent working family could afford to provide for itself.[20] It was generally agreed that standards were required for both quantity and quality.

The general practice was for prisoners to receive only bread and gruel for the first 14 days of imprisonment. Gruel at that time meant oatmeal boiled with water, or sometimes with a proportion of skimmed milk.[21] For the next 6 weeks, prisoners received meat and potatoes twice a week, and after that four times a week, with soup on the remaining days. Some critics said that it seemed irrational that those who had committed the most serious crimes, and therefore had the longest sentences, were fed the best. But prison doctors and administrators countered that continuing the more restricted diet would result in wasting and ill health, so that men would no longer be able to carry out the hard labor that was part of their sentence.[22]

18. Lawes (1853), pp. 539–40.
19. Lawes and Gilbert (1859), pp. 574–8.
20. Drummond and Wilbraham (1957), pp. 363–72; Tomlinson (1978); Johnston (1985).
21. Mayhew and Binny (1862), pp. 347–8.
22. Ibid., pp. 349–51.

Scurvy as a Symptom of Protein Deficiency

The supply of bread, potatoes and other vegetables varied slightly according to what was available at particular times. It had been concluded by 1843 that the numerous outbreaks of scurvy in English prisons in the previous 20 years had been associated with the omission of potatoes from the daily diet even when the total intake of albuminoids was normal.[23] A textbook of nutrition published in that year criticized Liebig's assertion in the first edition of his *Animal Chemistry* that albuminoids were the only true nutrients; it seemed clear also from the experience of victualing in the navy that nutritional requirements were more complex.[24]

In 1845 and 1846 northern Europe was hit by a devastating blight that destroyed the greater part of the potato crop. In Ireland, where the potato had become the major staple crop, the result was a terrible famine. In England and Scotland, where grains were still the major food crops and potatoes secondary, the significance of the antiscorbutic value of potatoes had not been appreciated, and there were widely distributed outbreaks of scurvy as a result of the blight.[25] A major outbreak at a prison in Scotland was investigated by Dr. Christison of Edinburgh University, remembered now for his work showing the danger of poisoning from lead pipes where the water supply was "soft."[26]

Christison had been convinced by Liebig's thesis that nitrogenous compounds were the only true nutrients. Yet the suspect prison diet supplied 135 g per day of albuminoids, an ample amount by either contemporary or modern standards. However, all but 23 g were in the form of gluten from bread. He therefore concluded that there were limitations to the extent that gluten could replace (or be converted to) albumin and fibrin. This was a small modification of the Liebig "paradigm." He explained that the successes reported from the use of "succulent vegetables" to treat scurvy were due to their content of "vegetable albumen."[27]

The argument was quickly refuted. As a physician from Glasgow pointed out, 4 oz. of lemon juice per day, which was known to prevent scurvy, contained only 2.5 g dry matter, and very little even of this small quantity was nitrogenous.[28]

In 1853 Lyon Playfair, who has already been referred to as a British protégé of Liebig, gave a lecture entitled "The Food of Man under Different Conditions of Age and Employment." He reported extensive

23. Baly (1843); Carpenter (1986a), pp. 99–102.
24. Pereira (1843), p. 47.
25. Carpenter (1986a), pp. 101–3.
26. Anonymous (1887b).
27. Christison (1847), pp. 888–90.
28. Anderson (1847), pp. 177–8.

calculations of the carbon and nitrogen contents of the diets furnished to paupers, prisoners and servicemen, and compared them with the diets of agricultural laborers. The average daily diet for paupers contained 14 g nitrogen and 230 g carbon, equivalent to approximately 88 g protein and 2,300 kcal per day in modern terms. Playfair said that no positive answer could yet be given to the question of how much "flesh-forming matter" (albuminoid) was required to support an adult man under normal conditions. It was determined by the amount of labor performed. He thought it appropriate therefore that prisoners sentenced to hard labor received a proportionally greater increase in "nitrogenous materials" than in fats or carbohydrates.[29]

The Work of Edward Smith

Another physician interested in nutritional questions during this period was Edward Smith.[30] In 1863 he had made the first survey, for the government, of the food eaten by low-income families.[31] Some of the results are summarized in Table 4.1 together with the calculated nutrient values of the diets, made more recently by a group at London University, and comparisons with the results of later surveys.[32]

Bread formed an extraordinarily large part of the ordinary diet in the 1860s. Even so, it had greater variety than the short-term prison diet, and Smith had argued that even short-term prisoners should be given a more substantial diet. Deterioration was admittedly not so marked in the short term, but the state should endeavor to return a prisoner to society in a strong rather than a weak condition, so that he could better compete for a job of honest labor and be less tempted to return to crime.[33]

Smith was also concerned that the "hard labor" that formed part of long sentences was unduly arduous.[34] The standard form of hard labor in London prisons consisted of continually climbing up the paddles of a large treadmill, rather like the side paddle wheels of a Victorian steamship. A row of them were linked together on the same axle and connected to a sail on the roof of the prison to provide resistance (Figure 4.2).[35] With the weight of the prisoner on a paddle, it fell, and he had to step up to the next one. He climbed in this way for 15 minutes, then rested for the same period while

29. Playfair (1853).
30. Chapman (1967); Barker, Oddy and Yudkin (1970); Carpenter (1991).
31. Smith (1864), p. 217.
32. Barker, Oddy, and Yudkin (1970), 35–49; U.S. Department of Agriculture (1983); U.S. Department of Agriculture (1985).
33. Smith (1857a).
34. Smith (1857b).
35. Mayhew and Binny (1862), pp. 303–7; Chapman (1967), pp. 5–9.

Table 4.1. *Mean results from Smith's 1863 survey of the diets of low-income indoor workers and their families compared with results from later surveys of relatively low income families*

		U.K.			U.S.
		1863	1933	1965	1977
Basic Victorian foods, per adult per week, oz.	Bread	145.6	65.6	40.0	25.6
	Potatoes	38.4	54.4	57.6	24.0
	Sugar	7.9	22.8	16.9	18.2
	Fats	4.7	10.2	8.3	11.5
	Meats	12.3	23.0	23.0	68.8
	Milk	16.0	36.0	80.0	90.0
Estimated nutrient intakes, per adult per day	Protein (g)	55	63	58	101
	Fat (g)	53	72	81	138
	Carbohydrates (g)	370	350	290	311
	Total energy (kcal)	2,190	2,320	2,040	2,880
	Calcium (mg)	360	370	820	1,030
	Iron (mg)	12.5[a]	8.0[b]	10.8	15.7

[a]During this period, significant additional iron was obtained from iron cooking pots and pans.
[b]During this period, there was some changeover from iron to aluminum cooking vessels.

another prisoner took his place. The standard day's labor consisted of 15 stints and corresponded to a climb of approximately 1.4 miles. Each prisoner did this 3 days per week, and the same food ration was provided each day regardless of the work done.

Smith set out to measure how arduous the work was by doing it himself while connected to an apparatus that he had developed for measuring the output of carbon dioxide from his lungs as the increase in weight of potassium hydroxide onto which it was absorbed, after the air had first been dried (Figure 4.3). He found that, on average, in the combined "15-minute work + 15-minute recovery period" he expired 19.6 g carbon per hour *more* than when resting. Since his 24-hour expiration when leading a restful life was altogether 223 g, the labor of a standard workday, that is, of 7½ hours "work + recovery," at this rate would have increased his estimated 24-hour output by 66%.[36]

Smith organized the collection of urine from a group of four prisoners for a 3-week period and analyzed it for the quantity of urea produced during

36. Smith (1859), pp. 692, 709–11.

Figure 4.2. Prisoners sentenced to hard labor at work on the treadmill at Brixton prison, London (British Register, August 1, 1823). (Watkinson Library, Trinity College, Hartford, Connecticut)

each day plus the following night. The overall average was 15.5 g nitrogen excreted in the form of urea. There was variation from day to day, but the average production on workdays was only 3% higher than on rest days.[37] This small, and not statistically significant, difference was in contrast to the very large increase in the carbon dioxide production of a man engaged in labor. Smith concluded that urea excretion was determined chiefly by the nitrogen content of the food and that the production of carbon dioxide was "the best measure we have of the vital functions attending muscular exertion."[38]

An Elegant Experiment in the Swiss Alps

Edward Smith did so much work, with thousands of analyses stretching over a period of 5 years, but he published it in such a way that some of

37. Smith (1862), p. 827.
38. Ibid., pp. 804–9, 831.

Figure 4.3. Edward Smith demonstrating the apparatus that he used to measure his excretion of carbon dioxide when working on the Brixton treadmill. (*Phil. Trans. Roy. Soc. London*, 1859)

the crucial points in his findings, which we have just described, were buried. However, two scientists in Switzerland saw the implications of Smith's work and, tying it in with a recent development in physics, were able in 1865, with no more than 4 days of experimental work and no subjects other than themselves, to write a paper that became one of the classics in the history of nutritional science.[39] Slightly the older of the two, at 35, was Adolf Fick, who had received his medical education in Germany and was now a professor of physiology at the University of Zurich. He had a long-standing interest in studying biological phenomena in terms of physics. His colleague, Johannes Wislicenus, was a 30-year-old associate professor in chemistry.[40]

Their experiment was designed to ask just one question: Does the amount of body substance broken down, and converted in part to urea, always provide enough energy to account for the energy expended in muscular exertion? It was Liebig's generalization that this was the only source of muscular energy, so that one well-documented exception should be enough to disprove it.

Fick and Wislicenus decided that, by ascending a mountain, they could

39. Fick and Wislicenus (1866); Coleman (1977), pp. 134–5.
40. Rothschuh (1971); Costa (1976).

establish a definite minimum amount of work, that is, by lifting their body weight through a definite height. There was a convenient mountain, the Faulhorn, on which they could ascend some 2,000 meters (m) by a steep path and rest overnight in a hotel at the top. Also, if they refrained from eating any albuminoids from noon on the day before their ascent until the following night, all the urea would have to come from their tissues rather than from any surplus food intake. They therefore prepared small cakes made from starch paste fried in plenty of fat and took along sugar dissolved in tea. There would also be some sugar in the beer and wine that they would drink en route. They would analyze their urine production for urea on the day of the ascent and reanalyze the samples for total nitrogen on returning to their laboratory.

They set out soon after 5 a.m. and reached the hotel at 1:20 p.m. They were enshrouded in a cold mist the whole time and did not become overheated, so that they did not consider losses by perspiration to have been a significant factor. They rested until 7 p.m., collecting "after-work urine," and then had a dinner consisting mainly of meat. On the next morning they collected their overnight urine and returned immediately to their laboratory to complete the analyses.

Their object was to compare the work done with the energy that could have been liberated from the breakdown of the muscle albuminoids. There was no doubt as to the external work achieved. Fick, with his hat, clothes and stick, weighed 66 kg and Wislicenus 76 kg. The endpoint of the climb was 1,956 m higher than the starting point. By simple multiplication, it could be shown that the external work done by the two men against the force of gravity was therefore 129,000 and 149,000 kg-m, respectively.

Then they used the concept of the conservation of energy, and its interconvertibility from one form to another, following the lead of J. R. Mayer as described in Chapter 3, but with more precise data. The work of James Joule and others had shown by then that the effort, or "energy," needed to raise a weight of 423 kg against the force of gravity by a height of 1 m was the same as that which, if all the effort were dissipated as friction, would give the heat required to raise 1 kg of water by 1°C (i.e., 1 kcal). Using this factor for the "mechanical equivalent of heat," one can say that the external work of Fick and Wislicenus was equivalent to 305 and 352 kcal, respectively.

Returning to the estimation of the energy that could have been obtained from the albuminoids, they found, as expected, that the value for the total nitrogen content of the urine was always slightly higher than that for urea nitrogen. To avoid controversy on this point, they used the higher values. Expressed as grams nitrogen per hour, the results can be summarized as follows:

Collection period	Fick	Wislicenus
The previous night	0.63	0.61
The morning of ascent (8 hours)	0.41	0.39
Rest after the ascent (6 hours)	0.40	0.40
The following night	0.45	0.51

The authors assumed that nitrogen from muscle which decomposed as a result of work could have appeared in the urine both during the work and in the following rest period. There was some variation in published analyses for the nitrogen content of muscle albuminoids, but no one had reported less than 15%. This conversion value would therefore give the highest possible weight of albuminoids metabolized for a given excretion of nitrogen. For the combined ascent and rest periods, the corresponding total weights of albuminoids metabolized became 37.2 and 37.0 g for Fick and Wislicenus, respectively. They then had to estimate the amount of energy that could have been released during the partial combustion of this quantity of material. Given the heats of combustion of carbon and hydrogen, they believed that the highest possible value would be 6.73 kcal/g albuminoid. On this basis the maximum energy obtainable would be 250 and 249 kcal, respectively.

These values were less than the external work accomplished by the two men, and no account had been taken of the internal work, such as the pumping of the heart and filling and emptying of the lungs, which the authors estimated to have been not less than another 30,000 kg-m per head. They therefore concluded that they had disproved Liebig's assertion. It was impossible for muscle substance to have provided the energy needed. Therefore, fats and carbohydrates must have provided the staple fuels for the muscles, since their metabolism was greatly increased during mechanical effort while that of the albuminoids was virtually unchanged. They suggested that the steam engine provided a useful analogy, since the iron machinery did not consume itself, but a separate fuel that was oxidized and provided the energy for the work indirectly through steam as an intermediate.

Later in the same year an Englishman, Edward Frankland, who was also Fick's brother-in-law, published an important supplement to the Swiss work.[41] His contribution was to measure directly the heat released by the complete combustion of a large number of foods and also of urea. He got them to burn by mixing them with potassium chlorate, and he did control runs with chlorate alone or mixed with substances whose heats of combustion were already known. He calculated the heat produced by immersing the combustion chamber in water and measuring the rise in temperature.

In this context Frankland's most important values were 5.10 kcal/g for dried beef muscle, from which the fat-soluble materials had been extracted

41. Frankland (1866a, b).

with ether, and 2.206 kcal/g for urea. Since 1 g albuminoids gave 0.33 g urea in metabolism, the remaining energy contained in the urea from 1 g albuminoids was 0.33 × 2.206 = 0.73 kcal. By difference, therefore, the energy released from metabolism of 1 g material was 5.10 − 0.73 = 4.37 kcal/g. This was considerably less than the indirect estimate of 6.73 kcal used by Fick and Wislicenus. Applying the revised figure, he found that the calories obtainable from the albuminoids that the two men metabolized were reduced to 163 and 162 kcal, equivalent to little more than half the external work performed.

Frankland also pointed out that all studies, both with mechanical systems and with isolated animal muscles, indicated that chemical energy was never realized entirely as mechanical work; it was accompanied by at least an equal amount of heat energy and usually more. In what he referred to as "Smith's highly important experiments on himself," the extra energy released from the body during treadmill work, as estimated from the carbon dioxide produced, was three to four times greater than the external work performed. The muscle, viewed as a machine, appeared to be operating at no more than 25 to 33% efficiency. This was a repetition of conclusions that Helmholtz had also drawn from Smith's data in a lecture given 5 years earlier.[42]

The actual calculations are not given in either paper, but we can reproduce them. As stated already, Smith expired an extra 19.6 g carbon per hour when he was working the treadmill "15 minutes on, 15 minutes off." Fat, carbohydrate and protein all yield approximately 10 kcal per gram carbon when oxidized, so that 19.6 g carbon is equivalent to 196 kcal. This in turn, using Joule's equivalent of 423 kg-m/kcal, comes to 82,900 kg-m. In an hour (i.e., with two work periods) Smith would have climbed 287 m. Since he weighed 89 kg, the external work done was 25,600 kg-m. This corresponds to 30% of the total energy estimated to have been liberated.

One group in Britain for whom these results were of immediate concern were army surgeons. They were responsible for ensuring that soldiers received rations that would allow them to engage in strenuous activity in wartime without becoming unduly fatigued. Professor Parkes of the Army School of Medicine was concerned to replicate the Swiss type of experiment and to check whether a 6-hour "after-work" period was enough to collect the protein breakdown products resulting from a period of labor. He organized a long series of trials with soldiers on both high-nitrogen and low-nitrogen diets, where days of long marches with full equipment were followed by rest periods of varying duration. He was finally convinced

42. Helmholtz (1861), pp. 355–7.

that "force necessary for great muscular work can be obtained by the muscles from fat and starch."[43]

Could the Liebig School Admit It Had Been Wrong?

In the 1850s Liebig's group expanded to include a physiologist, Theodor Bischoff. He was an admirer of Liebig and eager to extend and support his views of chemistry in relation to physiology. Later, he recruited an assistant, Carl Voit (in some cataloging systems, Karl Voit), who had been another admirer of Liebig as a medical student. They carried out very careful balance experiments with dogs, the results of which persuaded them that with their improved analytical methods they could obtain reliable nitrogen balance data and that little or no nitrogen was being lost through the lungs or skin.[44]

Bischoff and Voit then carried out trials with dogs fed meat, starch and fat in various proportions. They found, as had several other groups by then, that urea production was greater in dogs on high-nitrogen diets. Others had concluded that, although Liebig was obviously right in thinking that some dietary albuminoids went to muscle and other tissues to replace "broken-down" material, the excess, because of its rapid conversion to urea, must be directly metabolized in the blood or liver, with urea as one product, and that the remainder of the molecule was utilized for the same purposes as the starch and fat in the original diet.[45] But the Munich workers were unwilling to accept an idea so clearly contrary to Liebig's overall scheme. As an alternative, they produced a complicated argument to the effect that increasing the level of albuminoids in the diet resulted in the body having to increase the volume of blood plasma and of organs to hold them, and the heart having to do a proportionate amount of extra work. They ended by saying that "it is established for all time that the nitrogen-containing substances are the sources of physical power."[46]

With hindsight, we might rephrase that as: "Don't confuse me with facts; my mind is made up." Another German scientist wrote: "This scheme would resemble a miserably built machine which used all its resources to overcome its own internal resistance... and could be compared to the Danae, condemned by the Greek gods to spend an eternity trying to bail water out of a well with a sieve." He went on to say that "to the unprejudiced it would be immediately clear, from the observations of Bischoff and Voit, that the

43. Parkes (1871), p. 59.
44. Holmes (1987), pp. 251–7.
45. Lusk (1922), pp. 57–64.
46. Bischoff and Voit (1860).

breakdown of albuminoids has nothing to do with the muscle work of the organism . . . the only plausible explanation is that it occurs in a special organ, where it is activated by an enzyme [*ferment*], and everything indicates that this is the liver."[47]

Voit's "Work" Experiments

In Voit's next experiments he used a large exercise wheel, on the same principle as wheels employed in pet-mouse cages but large enough for 70-lb. dogs. On their running days in the wheel, Voit's dogs covered 16 km in 6 periods of 10 minutes each.[48] A typical set of results for 2 days in which a dog received water, but no food, was that on its running day it excreted 17.1 g urea and on its rest day 16.7 g. When it was being given 1,500 g lean meat per day, it excreted approximately 117 g urea on running days and 110 g on rest days.

The work involved in running and turning the wheel was calculated by Voit to be approximately 155,000 kg-m, equivalent to 366 kcal. This energy could be produced from the metabolism of approximately 85 g albuminoids, which would then yield 34 g urea. This was much greater than the total output of urea on fasting days, whether the dog was running or not, and vastly more than the 7-g increment on running days when the dog was eating 1,500 g lean meat.[49]

What Voit had expected to find was increased urea production as a result of work in sufficient quantity to account for the energy needed for the muscular work having come entirely from protein breakdown. This was not what he found. But he again thought of an argument that circumvented the obvious conclusion. He assumed that the dog on the days in which it was not fed was metabolizing its "flesh" and that this had the same composition as the "flesh" (i.e., meat) that he was feeding to the dog in the other periods. Then, assuming that the urinary nitrogen content reflected the nitrogen content of the flesh metabolized, and knowing the nitrogen content of the meat he fed to the dog, as well as an estimate of its caloric value, he could calculate the estimated total calories released from the dog's metabolism of its flesh. He reckoned that the dog that had been fasting, on a running day metabolized only 23 g flesh dry matter, which would yield 113 kcal. Since this was so much less than the external work calculated to have been equivalent to 366 kcal, he concluded that an animal's energy came not just from ordinary chemical reactions, but from intermolecular attractions and rearrangements that were not yet understood.[50]

47. Traube (1861), pp. 405–6.
48. Voit (1860), p. 154.
49. Ibid., pp. 156–72.
50. Ibid., pp. 190–1.

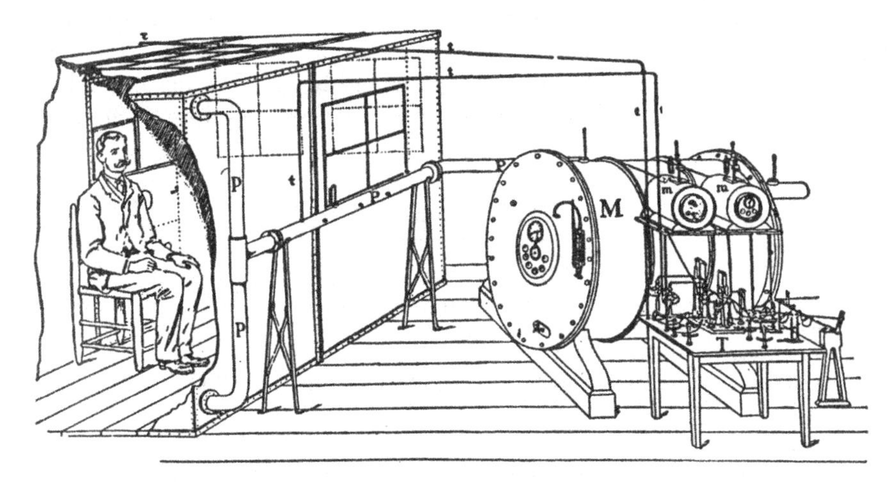

Figure 4.4. Cutaway sketch of Pettenkoffer's apparatus for measuring respiratory exchange in human subjects over a period of several days. (Bulletin 21, U.S. Department of Agriculture Office of Experiment Stations, 1895)

One would think that 17.1 g urea must have arisen from at least 43 g albuminoids, so that his "23 g flesh dry matter" is unintelligible. Also, since only nitrogen metabolism was measured, we have no way of knowing how much body fat may also have been metabolized. The modern reader would expect this to have been considerable. In contrast, the mechanical work was probably much less than Voit had calculated in terms of the dog moving in a series of kangaroo hops rather than running steadily up only a slight incline. This appears to have been another instance of avoiding an obvious conclusion at any cost.

Voit went on to enlist the help of Professor Pettenkoffer in designing and operating an apparatus containing a chamber in which a human subject could live for several days (Figure 4.4). The equipment included a means of measuring the oxygen absorbed in the chamber and the carbon dioxide and water vapor expired, in addition to the usual measurements of urinary nitrogen compounds.[51]

From our point of view, the most interesting comparisons were again for a human subject who, as in the preceding experiments with dogs, rested on some days and did physical work on others. The work consisted of giving a heavy crank some 7,500 turns by hand over a period of 9 hours. Unfortunately, the work done could not be quantified into meter-kilograms, but the subject said that he felt as tired from performing the task as if he had been on a long march.

51. Pettenkoffer and Voit (1868), pp. 472–8.

The experiments were repeated with the 70-kg subject receiving either nothing but water and a little extract of meat or else a mixed diet that included meat, eggs, bread, milk, fat, sugar and beer.[52] The mean results can be summarized as follows, with all the values expressed as grams per 24 hours:

		N out in urine			C out as CO_2 in breath	
	N in	Rest days	Work-days	C in	Rest days	Work-days
Hunger periods	1.5	12.4	12.3	3	203	333
On mixed diet	19.5	19.5	19.3	307	283	355

The development of the equipment was a considerable technical achievement, and the attention to detail was such that even a book taken into the chamber by a subject was weighed in and out to check whether it might have absorbed moisture. However, in relation to the principles of nutrition, it only confirmed that physical work was accompanied immediately by greatly increased oxidation of carbon, but resulted in little or no change in nitrogen metabolism.

The authors wrote that these results did not necessarily mean that the physical energy came from fat or sugar oxidation. Such a conclusion would require proving that not all of the energy released by oxidation appeared as heat – that is, that some was being diverted to work. They added that they were planning to develop equipment which would allow them to measure heat production directly while a subject was doing different kinds of work. But even if it turned out that oxidation of these other carbon compounds was providing the energy, the albuminoid muscle fibers must still be involved, since it was they that attracted the oxygen out of the blood and condensed it, and it was they that changed physically during muscle contraction.

The authors' hypothesis, though they realized it could only be tentative, was that the muscle albuminoids gradually, in periods of rest, built up a "tension" comparable to that in a wound-up spring, which could later be released as mechanical work by an act of free will, and that during this release oxygen was transferred from the albuminoids to fat, which oxidized with the release of heat. "The truth of this mysterious matter would have to wait for the outcome of further experiments."[53] This hypothesis did not, of course, explain the great increase in oxygen uptake during exercise.

52. Ibid., pp. 478–502.
53. Ibid., pp. 570–3.

Liebig's Last Interventions

Three years later, in 1870, Liebig returned to writing on the source of muscular energy, a subject that he had not touched for the previous 20 years, and two of his papers appeared in English translation in the following year. First, he acknowledged that he had himself been responsible for an error, namely, his assumption that urea production would always be proportional to work done. This had been disproved by Bischoff and Voit, though he claimed that he deserved a portion of the credit, since he had developed the improved method of analysis for urea that had given their work the required precision.[54]

Nevertheless, he still believed that the energy needed for muscular work came from albuminoids alone – not necessarily from their immediate combustion, but from internal molecular stresses gradually built up over a period of as long as 3 days. The spring of a wound-up clock could serve as an analogy. It was quite possible that the oxidative reactions occurring in the body could release more energy than those occurring in Frankland's combustion apparatus. It also seemed unreasonable that the daily loss of urea, even in someone on a nitrogen-free diet and at complete rest, should be so great if albuminoids were needed only to replace worn-out tissues and not as an energy source.[55]

In a third paper, after writing that he believed the formula of blood albumin to be $C_{216}H_{169}N_{27}S_2O_{68}$, he attacked his old disciple and colleague, Carl Voit. The latter's main offense had been to believe it possible that, under some circumstances, a portion of a high intake of albuminoids could be converted to stored fat. Liebig wrote, "In his [Voit's] hands facts are like wax, to which the wished-for form is given by kneading it," and he added that he was unhappy at the way in which, nowadays, people were drawing physiological conclusions before the necessary data had been sufficiently established.[56] This from Liebig of all people: what extraordinary impudence and a classic case of the pot calling the kettle black!

It was, understandably, too much for Voit, and he tried to set the record straight in a paper covering 99 pages. In it he pointed out where Liebig had adopted his (Voit's) views without acknowledgment, as in the concept of resting muscle fibers gaining tension like a wound-up spring. Then in his general conclusion he wrote: "He [Liebig] has brought about great progress through his ideas about the processes within the animal body but he has forgotten, to the sorrow of those who know and value his service to science

54. Liebig (1870a), p. 162.
55. Ibid., pp. 182–4; Liebig (1870b), pp. 77–8.
56. Liebig (1870c), p. 283.

better than do his flatterers, that these are all mere ideas and possibilities, whose validity has to be tested by actual animal studies."[57]

Liebig's Proprietary Products

We will return to the "muscle question" after a brief detour. In the period just under review, and for the remainder of the nineteenth century, the general public was to associate Liebig with two nutritional products that carried his name, but have not yet been described. Both were causes of contention between the famous chemist and other scientists and medical men.

One of these was Liebig's Infant Food, or Malt Soup (to translate the German name literally), which was advertised as a complete substitute for breast milk.[58] At the Medical Academy in Paris, speakers said that Liebig's great reputation as a chemist had created public interest in the product but that young infants did not thrive on it and the public must be warned about the danger of using it.[59] A Munich physician also reported adverse results.[60]

The other proprietary product was Liebig's Extract of Meat. Liebig wrote that it was prepared from finely chopped lean meat mixed with its own weight of cold water and slowly heated. After boiling for a minute or two, the liquid was decanted and strained. It could be used immediately for soup or boiled down to the consistency of honey, when it would keep indefinitely. (The commercial product was sold in the concentrated form.) Liebig concluded, "On the average, the soluble matters of 4 kg of flesh, after the coagulation of albumin and coloring matter, do not exceed 100 g and of this a very considerable proportion consists of organic salts, the phosphates being particularly abundant, while the remainder is formed of no less than five organic compounds." Creatine was the organic compound to which he devoted the most time. He found its formula to be $C_8N_3H_9O_4$ (which, with modern atomic weights, becomes $C_4N_3H_9O_2$), and showed that it readily decomposed to creatinine, which is excreted in the urine.[61]

Liebig put forward two arguments in support of the value of this product. The first was that, although plant albuminoids such as wheat gluten were fully equivalent to the albuminoids of muscle fibers, both our own muscles and the meat of animals contained, in addition, soluble or "juice" factors such as creatine, which were not found in plants. Our bodies therefore had to do extra work to convert plant material into these substances unless we either ate meat or supplemented plant foods with "extract of meat." These

57. Voit (1870), pp. 317, 399; Holmes (1964), p. cviii.
58. Liebig (1867).
59. Guibourt and Depaul (1867).
60. Poppel (1869).
61. Liebig (1847), pp. 39–60, 131–2.

substances were true nutrients because they were identical with compounds produced in the body from albuminoids.[62]

He offered a second argument in reply to critics who complained that the public was being misled by commercial advertising of the product, which seemed to imply that 1 oz. of concentrated extract was the *equivalent* of 2 lb. of meat and that invalids in particular would be strengthened by it when they were unable to eat solid food. For example, Edward Smith had said at a public meeting in England in 1872 that the product had been sold in large quantities since Liebig had lent his name to it, but it had little nutritional value; the flavor of meat disguised the real poverty of the substance. However, with hot water it made a stimulating drink that was comparable to tea or coffee.[63]

These views were summarized in the London *Times*, which published a reply from Liebig a few weeks later. In his unique style, he characterized Smith's views as comical and "without even a faint notion of the science of chemistry," but then agreed with Smith that "extract of meat is not nutriment in the ordinary sense.... Like tea, it possesses a far higher importance by certain medical properties of a peculiar kind.... taken in proper proportions it strengthens the internal resistance of the body to the most various external injurious influences."[64] In the following year he also wrote that meat extract increased the working capacity of the body through its effect on the nerves; he added that people did not judge foods by their carbon and nitrogen contents, but paid more for meat (than for liver or cheese) because it contained certain other substances.[65]

The claims for extract of meat having special qualities have never been substantiated.

Summing Up

By the end of the 30-year period covered in this chapter most of the beliefs held at the beginning had disintegrated. On the purely chemical side, differences were found in the carbon–nitrogen ratio of different albuminoids, and the existence of Mulder's "protein radical" could not be confirmed. This meant that animal proteins had, at least to some extent, to be manufactured by animals themselves, which was contrary to the earlier doctrine.

Experiments with both humans and animals also showed that physical work was not necessarily accompanied by a significant breakdown of tissue protein and increased excretion of urea, though it was always accompanied

62. Liebig (1870b), pp. 108, 114.
63. Smith (1874), pp. 87–9; Finlay (1992), pp. 411–13.
64. Liebig (1872).
65. Liebig (1873).

by increased oxidation of carbon, which must have come from the body's stores of fat and/or carbohydrate. By 1870 the matter had been settled for most of those interested, though the German school continued to try to find arguments that circumvented the obvious conclusion. But by the end of the century reviewers were unanimous in concluding that protein was not the main, or obligatory, source of energy.[66]

Now that Mulder's concept is well in the past, we can use the word "protein" again, in the modern sense of any type of "albuminoid." The development of knowledge as to how muscles actually worked was slow and extremely complex, and is the subject of a comprehensive work.[67] In some aspects, Liebig's picture was correct. The protein fibers do store energy, and creatine does serve as an intermediate energy store involved in the process of replenishment, but the processes are fully reversible, and it is the oxidation of carbohydrates and fatty acids that we rely on for continued physical work.

One last point of terminology: It is now becoming standard practice in scientific publications to replace the kilocalorie as a measure of energy expenditure with the joule (J). One calorie is usually taken as equivalent to 4.19 J. For example, what used to be described as a daily energy expenditure of 2,000 kcal/day now becomes 8.38 megajoules (MJ). In trying to connect this relationship to James Joule's own conversion factor, one must remember that the force of gravity at the earth's surface is approximately equivalent to 10 m/sec^2, and the present unit is the work required to move a mass of 1 kg a distance of 1 m against a force equivalent to 1 m/sec^2.

66. Paton (1895); Armsby (1906), pp. 194–209.
67. Needham (1971).

5 Vegetarian Philosophies and Voit's Standards, 1875–1893

In the preceding chapters we followed the gradual development in Europe of scientific knowledge about the chemical composition of foods and our requirements for different types of nutrients. The writers considered up to this point judged the suitability of different foods – meats, bread, fruit and so on – entirely in terms of their digestibility and their capacity to meet the needs of tissue growth or replacement.

The Romantics' Abhorrence of Killing

From the beginning of the nineteenth century, another school of thought developed in England and then in the United States. Its central belief was that it was morally wrong to kill animals in order to eat them. This was not an entirely novel idea. Many early Greek philosophers had had the same belief, as had followers of several Eastern religions. Some earlier Europeans, such as Leonardo da Vinci, had also been vegetarians. However, there had been no continuing tradition of vegetarianism in the West.[1] The Catholic church forbade the eating of meat on fast days, but by implication considered it acceptable on other days. And the Old Testament contained references to Jehovah's pleasure at the sacrifice of animals on his altars.

The reappearance of the vegetarian ideal in the 1790s and early 1800s seems to have been related to the Romantic movement, which began as a reaction against the formalism and arid rationality of the eighteenth-century Age of Enlightenment in France. Jean-Jacques Rousseau had attacked the artificiality of French society and argued that a simple kindness of heart was more important than sophisticated knowledge.

One of the first Englishmen to write on vegetarianism in this period quoted extensively from Rousseau – for example, "The only way in which man can be happy is to be mild, benevolent and humane, in contrast to the

1. Barkas (1975), pp. 69–73; Dombrowski (1984).

present bloodshed and cruelty." The author, Joseph Ritson, argued that humans had an animal body, so that the killing of other animals was halfway cannibalism. Moreover, it did not appear from the anatomy of our teeth and digestive tract that we were intended to be meat eaters. The animal closest to humans in the arrangement of its teeth, the orangutan, was a vegetarian in its natural environment. Ritson cited examples of peoples with a reputation for strength and good health who lived without meat. Among these were Irish laborers, whose diet consisted virtually entirely of potatoes and buttermilk. Barbarous killing to obtain meat was therefore not essential and done only "to gratify luxury".[2]

Nine years later, John Newton published *The Return to Nature,* in which he argued that "man, of all races, is the most diseased." This was because he had "quit the nutriment on which alone Nature had destined him to enjoy a state of perfect health." Newton personally knew no less than 25 vegetarians and all were in excellent health. "In the happy state of innocence [i.e., in the Garden of Eden], man was commanded to eat, not fish or meat, but the fruits of the Earth of every sort."[3] Certainly, if the Creator desired humanity to follow a moral order, and if He regarded killing other animals as immoral, it would seem to follow that He would also have designed our own system so that optimal health could be maintained without our having to kill.

The poet Shelley, after only one year as an undergraduate at Oxford, was expelled for writing a pamphlet in defense of atheism. He became a friend of John Newton and was converted to vegetarianism. When still only 21, he published a pamphlet on the subject, claiming, like his predecessors, that "the present depravity of the physical and moral nature of man originated in his unnatural habits of life." He added that he believed that fire and cooking had been used to disguise man's horror of eating bits of bleeding carcass, and "from this moment his vitals were devoured by the vultures of disease." Like Newton he assumed that "man, and the animals whom he has depraved by his dominion, are alone diseased." Except in children, no instinct remained for what was natural.[4] In arresting terms, he wrote:

> The whole of human science is comprised in one question: how can the advantages of intellect and civilization be reconciled with the liberty and pure pleasures of natural life; how can we take the benefits and reject the evils of the system which is now interwoven with the fiber of our being?[5]

These ideas were, of course, entirely at odds with both the orthodox science of the time and the views of "the man in the street." As we saw in

2. Ritson (1802), pp. 40, 44, 78.
3. Newton (1811), pp. 63, 65, 109.
4. Shelley (1813), pp. 9–11.
5. Ibid., p. 12.

Chapter 3, the ideal food was considered to be one with the composition of animal tissues, and plant foods were judged by how nearly they approached that ideal in their nitrogen content. The popular feeling in England during this period of the Napoleonic Wars, and one that contributed to morale in the armed services, was that the relatively high meat diet in Britain, as compared with the diet in France, made "one Englishman equivalent to at least two Frenchmen." The complications of French cooking were regarded contemptuously as being merely ways of "making a little meat go a long way."[6]

The first preacher to urge vegetarianism on members of his congregation appears to have been the Reverend William Cowherd, founder of the independent Bible Christian Church, in Lancashire in 1807.[7] Cowherd and his congregation were said to have become unpopular in the area because of their criticisms of their neighbors' habits. One of Cowherd's assistants, William Metcalfe, emigrated to Philadelphia with some of the flock in 1817 and made further converts there.

The new environment was probably more favorable for these immigrants. There was a mood of optimism in the United States at this time and a feeling among ordinary people that there was no limit to their opportunity for a good life if they worked hard and abstained from every kind of personal excess. Many writers urged, in particular, that ill health and epidemics not be accepted as trials sent by God to be endured as a means of acquiring merit in His eyes. Health was a duty and sickness a sin, reflecting gluttony and immoderate living.[8] One writer has said that during this period "perfectionism swept across denominational barriers and penetrated even secular thought. . . . The progress of the country suddenly seemed to depend upon the regeneration of the individual and the contagion of example."[9]

Sylvester Graham

In 1846 Metcalfe wrote that he had one hundred people who had been abstaining from animal foods and intoxicating drinks. The convert who was eventually to be the most famous was Sylvester Graham. Since it has been suggested that Graham's early experiences explain some of his later teachings, I will summarize these briefly. He was born in 1794 as the youngest of a large family. His father, a retired Connecticut clergyman, died two years later, and soon afterward his mother lost her mind. He was cared for by a series of relatives, his education interrupted by bouts of ill health.

6. Drummond and Wilbraham (1939), p. 254.
7. Forward (1898), p. 7; Carson (1957), pp. 15–17.
8. Fowler (1851), p. 49; Whorton (1982), pp. 13–37.
9. Thomas (1965), pp. 658–9.

Finally, at the age of 30 he became a Presbyterian preacher and, after private tutoring, a regular minister in New Jersey. He began to purchase books on physiology, anatomy and nutrition. In 1830 the Philadelphia Temperance Society appointed him as a lecturer, and he began to speak not only on the ill effects of alcohol, but also on the importance of chastity, the prevention of cholera and diet reform, including the elimination of meat.

Graham was a very successful lecturer and spoke in New York, Boston and Philadelphia for fees as high as $300 per night, at a time when a nurse might earn $3 per week. However, protesters attempted on several occasions to stop his lectures. Some were butchers and others people protesting his discussion of sexual subjects. In 1837 he arranged for the publication of the *Graham Journal of Health and Longevity,* and in several cities Graham boardinghouses were set up where his recommendations could be followed. In 1839 he published a large work, *The Science of Human Life,* and then went into semiretirement. He died in 1851 at the age of 57. His aggressive and impassioned style made him a successful public speaker, but he was apparently aloof and crotchety toward his admirers.[10]

Graham is more complex and interesting than one might expect the originator of "Graham crackers" to be. First, he was not a straightforward puritan, suspicious of any kind of pleasure. For example, he wrote:

> Exercise is important, but it must be enjoyed. Religious prejudice against dancing is altogether ill-founded.... better that people should come together to sing and dance, in the health and exhilaration of their spirits than ... that they should endure a miserable existence in moping melancholy.... Children require much exercise in the open air. Action is as instinctively natural to children as breathing, and it is unnatural and improper to restrain them from it ... for any considerable time. Girls should be allowed as much freedom of action, in childhood, as boys.[11]

Graham's philosophy began with a consideration of human nature –

> which has occupied the minds of so many thinkers.... In the human body, matter and vitality, and mind and moral feelings are mysteriously associated.... By this wonderful union of intellectual and moral powers with organized matter, man alone, of all terrestrial beings, is brought into a two-fold relation with his Creator.... The animal nature of man may be considered as the basis of human existence. Its passions and desires ... constitute the primary and principal elements of activity to his mental powers, and tend continually to cause his rationality to concur with his animal indulgence.... Man, unlike animals, has both the voluntary and intellectual powers and the natural propensity to violate the constitutional laws of his nature, and thus deprive, deteriorate and destroy himself.[12]

10. Naylor (1942); Nissenbaum (1980), pp. 9–18; Hammond (1987), pp. 43–53.
11. Graham (1839), Vol. 2, pp. 119, 655–6.
12. Ibid., Vol. 1, pp. 424–5, 428.

He added that an addiction could "establish in the body an appetite whose despotic and often irresistible influence... compels the understanding and will to comply with its demands."[13] The obvious addiction that would have sprung to the reader's mind at that time was alcoholism, but Graham went on to develop a broader argument. History had repeatedly shown that "savage" peoples could work their way up to a life of civilization and luxury, only to become debilitated and effete and to fall back again into poverty and servitude.

Disease Due to "Overstimulation"

Graham's ideas as to the kind of life that would keep people vigorous and free of disease were in line with some of the orthodox medical ideas of the period. Doctors believed that the body was healthiest and had the greatest resistance to disease when in balance between the two extremes of weakness, or debility, on the one hand, and "plethora" (i.e., ruddiness with excess of blood and overheating), on the other. The purpose of their practice was therefore to bring their patients back to the ideal mean position, in part by prescribing the foods they could eat. Meat, eggs and spices were all "heating," as were coffee, tea and alcohol, and were therefore denied to plethoric patients; they were to receive fruit and vegetables, regarded as "cooling." Pregnant women, for example, were regarded as susceptible to plethora, and one aristocratic lady expecting her first child in 1810 wrote to a friend that, because of her physician's instructions, "I am now living exactly like a horse on grass food and water."[14]

Benjamin Rush, the doyen of the Philadelphia Medical School at the beginning of the century, had taught that the direct cause of disease was "morbid excitement," that is, excessive stimulation or irritation, either physical or mental, operating on a body already debilitated, either indirectly by previous overstimulation or directly by cold, hunger, grief, and so on. The main route of exterior physical stimulation was the gastrointestinal tract, and overeating, especially of spiced foods, was therefore a major source of overstimulation of the body.[15]

In 1826 a translation of the French physiologist Broussais's similar views on overstimulation as the cause of debility and disease had been published in Philadelphia. He wrote: "Strong and black meats, full of extractive matter ... high seasoned and fermented liquors have the double bad effect of supplying an abundant and substantive chyle, and of exercising [i.e., straining] the assimilative power of the stomach.... While the man is yet young he

13. Ibid., pp. 437–8.
14. Lewis (1986), pp. 130–1.
15. Nissenbaum (1980), pp. 54–7.

resists for a length of time such excesses. But there is a limit to everything. . . . [Finally] the fundamental organs, after having resisted acute congestions, sink under chronic irritations."[16]

Graham followed these precepts in teaching that people should eat as little food as they found to be consistent with maintaining their energy and that they should cut out all meat and stimulants, including alcohol, tea, coffee, pepper and mustard. He also implied, at one point, that they should eat foods that were easily digested.[17] However, William Beaumont, in his famous work with a patient who had a wound that provided direct access to his stomach, had shown that meat was "digested" (i.e., dissolved) there more quickly than were vegetable foods.[18]

Graham and other vegetarian writers accepted Beaumont's finding, but claimed that the stimulation and "heat" felt by people after having a meat meal was evidence that their rapid digestion required a large expenditure of vital energy.[19] (The modern scientist would agree that the digestion and absorption of high-protein foods such as meat result in an increase in metabolic rate. This is attributed principally to the complex, energy-using chemical reactions involved in handling the absorbed digestion products.)

With regard to vegetable food, Graham argued that the slower digestion was gentle and unstimulating, and allowed people to work for a longer period, without the need for a further meal. He accepted that a portion of vegetable foods remained indigestible, but said that the structure of the gut showed that it was designed to deal with a certain amount of "innutritious" matter (i.e., fiber, in modern terminology) and that this was "necessary to sustain the functional powers of the alimentary organs." He used this point in criticizing the conclusions drawn from Magendie's experiments, in which dogs died when fed pure sugar or white bread. These did not prove the need for a high intake of nitrogen, since no diet lacking non-nutritious matter could be expected to keep an animal healthy. He was, in any case, critical of attempts to characterize diets by their proportions of carbon and nitrogen, because living beings had vital powers that did not obey the law of dead chemistry.[20]

Graham argued that wheat bread was a particularly good food. He did not seem to regard the leavening with yeast and subsequent baking as "unnatural." He did, however, disapprove of commercially prepared bread because of the chemicals added to it.[21] (There is evidence that this was a real problem in that period.) He also argued against the fractionation of

16. Broussais (1826), pp. 325–6.
17. Graham (1839), Vol. 1, p. 540.
18. Beaumont (1833), p. 46.
19. Graham (1839), Vol. 2, pp. 108–12; Whorton (1982), pp. 81–2.
20. Graham (1839), Vol. 1, pp. 542–4; Vol. 2, p. 116.
21. Ibid. Vol. 2, pp. 423–5.

ground wheat to separate the white flour from the branny fraction. White bread was too concentrated and constipating. Bran was non-nutritious, but it was "not a mere mechanical irritant"; it corrected chronic diarrhea as well as constipation, "probably from the soothing action of the mucilage."[22]

Whole-meal bread seemed to have a symbolic importance for Graham. It should be baked by the mother of a family – not by servants – and her love for her family would ensure that the process received the care it deserved.[23] The bread, made with clean wheat, freshly ground, would be the centerpiece for the family coming together at their dinner table. This was a romantic picture painted by someone whose own mother had been unable to carry out this role.

Diet and Sex

To the modern reader it seems strange that health reformers in the 1830s should have considered even modest stimulation a danger.[24] Perhaps the greatest danger in their eyes was that such stimulation or irritation would be transmitted through nervous connections to the reproductive system. Earlier writers had referred to the exhausting effects of loss of semen – each emission resulting in a shortening of life by so many days, or "the loss of 1 ounce of this rectified spirit being more weakening than the loss of 40 ounces of blood."[25]

However, this was not Graham's main concern; his strictures applied to either sex. In his lectures on chastity he said that "the convulsive paroxysms attending venereal indulgence... cause the most powerful agitation to the whole system that it is ever subject to.... The powerfully excited and convulsed heart drives the blood, in fearful congestion to the principal viscera – producing depression... rupture, inflammation and sometimes disorganization."[26] Apparently, orgasm, or even "the actual exercise of the genital organs," was not required for this dreadful sequence of events. "Day dreams and amorous reveries" or even "an aching sensibility" could themselves lead eventually to debility, disease and death.[27]

What shocked people at the time about Graham's views was that intercourse between man and wife was made to seem as undesirable as adultery, since he was arguing on hygienic grounds rather than moral ones. To counter this point, Graham could only say that sex in marriage was less damaging

22. Ibid., pp. 428–30.
23. Ibid., pp. 454–67.
24. Whorton (1982), p. 78.
25. Sokolow (1983), p. 79.
26. Quoted by Whorton (1982), p. 93.
27. Quoted by Nissenbaum (1980), p. 113.

than adultery because, in general, it was less exciting; but even vigorous young couples should indulge no more than once a month.[28]

Why Graham, and others, in this period should have professed such views about sex has been discussed elsewhere.[29] It is more relevant, in the present context, to consider what evidence he had for associating meat eating with lust. Benjamin Rush, writing on diseases of the mind in 1812, referred to cases of excessive sexual appetite and their treatment. He believed that one remote cause was "excessive eating, more especially of highly seasoned animal food," and that treatment should include "a diet, consisting simply of vegetables, without any of the condiments that are taken with them." He added that Dr. Stark, who had reported the results of his own dietary experiments, "found his venereal desires almost extinguished by living upon bread and water."[30] This was perhaps not surprising, since he died as a result of this self-experimentation, probably of scurvy.[31] Graham himself did not cite Rush or offer specific evidence for his belief.

Graham's lectures were well attended and his books sold in large numbers but his followers formed only a tiny portion of the U.S. population. The more common view was that vegetarianism was inconsistent with physical vigor (Figure 5.1). Graham's type of teaching was subjected to some witty attacks:

> Here is an exact portrait of a modern lecturer on Dietetics . . . general ignorance with a smattering of medical knowledge; . . . all the great objects of amelioration and amendment are to be accomplished by the substitution of unbolted flour in the place of pure wheat and solid animal food. . . . man has no imagination, no heart, no soul . . . but is all stomach. . . . Vegetable food is said to preserve a delicacy of feeling seldom enjoyed by those who live principally on meat. . . . Green peas and cabbage are become meta-physical.[32]

John Harvey Kellogg

Other proponents of vegetarianism continued to lecture after Graham's death in 1851, and one religious group adopted his teaching virtually in toto. These were the Seventh Day Adventists, so called because they believed that Saturday, the seventh day of the week, should be set aside for worship and rest, rather than Sunday, and that the Advent, or Second Coming of Christ, would be in the near future. They started as three small congregations

28. Ibid., pp. 118–19.
29. Sokolow (1983), pp. 91–5.
30. Rush (1812), pp. 350–3.
31. Carpenter (1986a), p. 201.
32. Anonymous (1837), pp. 337, 342, 350–1.

Figure 5.1. Caricature exemplifying the common view that vegetarians necessarily became weakly. Members of the Vegetarian Society are portrayed as having to be carried to their annual banquet on stretchers. (*Le Charivari*, Paris, 1853)

in New England in 1845. By 1855 they had moved their headquarters to Battle Creek, Michigan.[33]

So that church members could maintain healthy living in preparation for the Advent, prohibitions were gradually imposed – on alcohol, tobacco, tea, coffee, pepper and other condiments and finally meat in the 1860s. In 1866 the Western Health Reform Institute was established at Battle Creek for the purpose of teaching and bringing invalid Adventists back to fitness through "healthful living" in a religious atmosphere, rather than through the use of drugs. At first those in charge had no medical qualifications, but in 1876, when the institute's affairs were languishing, John Harvey Kellogg, 24 years old, was appointed as physician in chief. Raised as an Adventist, he had been converted to Grahamism as a boy, and in 1872 the organization had sent him to New York to obtain an orthodox medical education. On his return he received permission to create a much larger medical sanitarium.

33. Land (1986), pp. 39–51.

Although this facility attracted patients from all over the country, the church organization was unhappy because Kellogg appeared to be putting health reform activities ahead of religion, placing too much emphasis on bringing in celebrities such as Henry Ford and John Rockefeller, and providing entertainment with a resident string orchestra in the dining room.[34]

Kellogg lived until 1943 and wrote many books, which continued to appear in new editions in the 1920s and 1930s and made him wealthy. However, his main ideas were set in place within a few years of his taking charge of the sanitarium. On the damaging effects of sex his views were at least as strong as those of Graham. He was particularly concerned as to "the almost universal vice of self-abuse among the young"; "much of the nervousness, hysteria and general worthlessness of the girls of the rising generation originates in this cause alone."[35] "Meat of all kinds . . . is stimulating and should not be freely used by anyone whose nervous system is already over-excited and irritable."[36]

He believed that use of the generative organs caused a nervous shock to the whole system and that unnecessarily frequent intercourse in marriage could lead to insanity. Although he himself married, he said that he wanted to show that continence was still possible; he and his wife had separate rooms and adopted children but had none of their own.[37]

Autointoxication

In his writing on diet Kellogg referred not only to the dangers of general "overstimulation" from meat eating, but also to the poisoning and the individual diseases associated with the practice. In particular he had collected data suggesting that there was a lower incidence of cancer among vegetarians.[38]

Unlike Graham, he believed that the characteristic effects of foods could be explained in terms of their chemical components. The bad effects of meat he attributed to its high protein content.[39] In the main, these effects came indirectly from the action of bacteria, whose significance had become apparent only after Graham's time. It was now realized that these organisms were responsible for the phenomenon discussed in earlier chapters – namely, that, in moist storage, foods high in nitrogen putrefied and went alkaline, while high-carbohydrate foods fermented, with the production of acid. The chemical products of the acid fermentation were generally harmless, but the

34. Carson (1957), pp. 88–103; Whorton (1982), pp. 202–3; Land (1986), pp. 69–70, 133–8.
35. Kellogg (1893), pp. 144, 148.
36. Ibid., p. 221.
37. Sokolow (1983), p. 162; Carson (1957), pp. 110–11; Money (1985), p. 84.
38. Kellogg (1877), p. 55; Kellogg (1921), p. 491; Kellogg (1923), pp. 145–6, 253–5.
39. Kellogg (1922), pp. 35–7.

bacterial putrefaction of proteins gave some products that not only smelled repulsive but were poisonous. In fact, given the favorable light in which nitrogen-containing materials were considered in earlier contexts, the associations that Kellogg made come as something of a shock:

> The offensive odor arising from the burning of protein is due to the poisonous compounds which are formed from nitrogen and sulphur. Compounds of these substances, formed either in the chemical laboratory or in the mysterious laboratory of the plant, are among the most poisonous known to man. Nitrogen is an essential constituent of the high explosives used in warfare. It enters into the venoms of snakes and the virulent poisons produced by bacteria.[40]

It was believed by many physicians in the 1880s that this putrefaction could occur in the human colon, or large intestine, especially when someone had eaten more protein than could be completely digested in the small intestine. The result would be *autointoxication* from the toxic products.[41] It was further supposed that bacteria of the putrefactive type had reached the colon originally from meat that was already infected. Kellogg was convinced of the importance of this concept and added that constipation – the disease of civilization resulting from overrefined foods – gave the bacteria more time to do their deadly work. To ensure that food residues did not stagnate in the colon, one should keep to a diet that promoted three bowel movements per day, was free of meat and was low in total protein content.[42] Obviously, cereals, fruit and vegetables were desirable types of food. Kellogg and his wife developed recipes for precooked, ready-to-eat wheat and oat granules that they called "granola." His brother William then developed flaking procedures on an industrial scale. Wheat flakes were officially introduced at the Seventh Day Adventists' annual meeting in Battle Creek in 1895, and by 1906 a method for producing corn flakes was also perfected.[43]

Today, few of the millions who eat the Kellogg Company's breakfast cereals know of the beliefs of their inventors, and if they did would probably consider them unbalanced. However, to jump forward in time for a moment, it is interesting that the Seventh Day Adventists are a thriving church and that modern studies of the health records of members following the traditional prohibitions have shown that they do indeed have a lower risk of developing cancer than their neighbors who have no such prohibitions.[44]

It is also of interest that, as visitors to the West End shopping area of London know, from at least 1973 to 1990 an anonymous man has consistently walked the streets carrying a banner proclaiming, "Less lust from less

40. Kellogg (1921), p. 31.
41. Bouchard (1894), pp. 138–53; Hudson (1989), pp. 396–400.
42. Kellogg (1923), pp. 632–3.
43. Money (1985), pp. 24–6.
44. Berkel and de Waard (1983).

Figure 5.2. The "protein man" (1987), a familiar figure in London's central shopping district for many years. (Art Director's Photo Library, London)

protein" and selling a pamphlet now in its 35th edition that gives details of his ideas (Figure 5.2). But one must wonder whether some people who have seen his banner have been inclined to change their diet in the opposite direction to the one he intended.

The Opinions of the Establishment

We left Chapter 4 with scientists' general acceptance by the 1870s that dietary protein was not the essential energy source for muscular contraction.

The uses of protein for the growth of new tissue and the replacement of existing, worn-out tissue were not contested.

Carl Voit, who had attacked Liebig in his "25-year review" in 1870 (as discussed in Chapter 4), continued to organize a long series of studies on protein nutrition in Munich and became the new authority, though not an undisputed one. In 1875, in a lecture to people with responsibility for providing food in different kinds of institutions, he proposed a daily dietary standard for an average worker. This consisted of 118 g protein, 56 g fat and 500 g carbohydrate, with the proviso that one-half of the protein be from animal sources. The quantities of fat and carbohydrate could also be modified as long as they included no more than 500 g carbohydrate and their combined carbon content remained the same. Later he was to modify this by saying that their combined energy value should remain the same.[45]

Voit emphasized that his standard did not apply to the average man, but to a physical worker, that is, a necessarily well-muscled person. Inactive people, like senior citizens in a nursing home or confined prisoners, needed less. He understood that at first sight this did not seem to be consistent with his own experiments showing that protein metabolism, as judged by the excretion of urea, did not increase as a result of physical activity. However, the protein requirement was proportional to muscle mass, and the ability to do heavy work required the maintenance of a larger mass. Everyone accepted that a large draft horse at rest needed more dietary protein than a pony did. In the same way the physical worker needed to maintain his protein consumption even on his rest days so that he would not lose strength before returning to work.[46]

Voit accepted the criticism that his standard was based largely on what men ate rather than what they needed to eat, but in 1881 he published an expanded explanation of his standard and referred to another publication by a visiting worker in his laboratory, claiming that it provided data confirming the need for 118 g protein (equivalent to 105 g *digestible* protein) by the average worker.[47] The author was Hamilton C. Bowie, from San Francisco, who was in Munich in 1878–9. Bowie had begun by citing papers of scientists who had carried out studies on themselves and found that they maintained nitrogen balance on diets that contained 100 g protein or less. But he pointed out that these people did not even have to remain standing all day. Voit's "average worker" was a man who could maintain physical work for 10 hours per day at a trade such as carpentry or bricklaying – that is, something more strenuous than tailoring, but less active than work-

45. Voit (1876), pp. 21–2; Voit (1889), p. 243.
46. Voit (1876), pp. 23–4.
47. Voit (1881), p. 525.

ing as a blacksmith. There was no contradiction, therefore, in scientists, with their smaller muscle mass, requiring less protein.[48]

Bowie's own calculations were based on his reanalysis of data that had been obtained with a laboratory assistant (P.P.), 43 years old, weighing 74 kg and described as "big and muscular." His work included cleaning glassware and keeping the fire going in the laboratory oven, and his normal diet included cheese, sausage, meat dumplings, bread and beer. In one 24-hour period his urine was found to contain 18.5 g nitrogen, and in another 14.4 g. Assuming that he also had a typical loss of 2.3 g nitrogen per day in his feces, the average combined loss per day was 18.75 g. Bowie assumed that he was in equilibrium (i.e., neither gaining nor losing body mass) at this time and that his average daily diet must therefore also contain 18.75 g nitrogen. Taking a factor of 6.45 for converting nitrogen to protein, this corresponded to an average of approximately 120 g protein per day.[49]

Max Rubner, another colleague in the Munich laboratory, had published the first results of the work with this subject, which had been designed to measure the digestibility of individual foods or simple combinations. For each study the subject consumed on the first day nothing but about 2 liters of milk. On the next 2 days (or occasionally 3 days) he ate the test diet, and finally spent the last day with nothing but another 2 liters of milk. The "milk days" were used to yield fecal material with a different character. The analysts could then separate for analysis the fecal material appearing between the "milky" ones. This material could be assumed to have originated from the test diet even though it appeared a day or so later.[50] Altogether 10 such studies were carried out using subject P.P. (who was recoded "D" in this paper). Rubner had also reported the nitrogen contents of the urine produced on the test days, although he made no further use of these results.[51]

Bowie calculated the nitrogen balance in each test period by subtracting from the nitrogen present in the test diet over the 2 (or 3) days both the corresponding fecal nitrogen and the urinary nitrogen during the test period.[52] His findings are summarized in Table 5.1. In the original paper he also calculated balances for some of the 1-day periods on milk, but the results were erratic, and I have omitted them. Figure 5.3 shows graphically the relation between the amount of "digestible protein" eaten and the nitrogen balance of the subject (i.e., the balance of nitrogen consumed minus nitrogen excreted).

The data show that, in dietary periods when he ate less than 65 g digestible

48. Bowie (1879), pp. 466–71, 475.
49. Ibid., p. 474.
50. Rubner (1879), pp. 119–20.
51. Ibid., pp. 138–77.
52. Bowie (1879), p. 479.

Table 5.1. *Results of short-term nitrogen balance experiments in a human subject (g/day)*

	Primary data			Calculated data	
Food	N in	Fecal N	Urine N	N balance	Digestible protein[a]
White bread, yeast[b]	7.6	2.0	11.2	−5.6	36.1
Macaroni noodles	10.9	1.9	16.0	−7.0	58.0
Black bread	13.3	4.3	12.5	−3.5	58.3
Dumplings[b]	11.9	21.3	13.8	−4.2	61.9
White bread[b]	13.0	2.4	12.6	−2.0	64.9
Macaroni, gluten	22.6	2.5	18.0	+2.1	129.5
Meat, bread, lard (200 g)	23.5	3.3	18.3	+1.9	130.3
Milk, bread, butter (240 g)	23.0	2.6	16.2	+4.2	131.6
Meat, bread, lard (100 g)	23.7	2.9	23.5	−2.7	133.5
Meat, bread, fats (380 g)	23.4	2.1	15.5	+5.8	137.1

[a]The "digested crude protein" was calculated as ("N in" minus "fecal N") × 6.45.
[b]These test foods were given for 3 days, the remainder for 2 days.
Sources: The primary data are from Rubner (1879) and the calculated data (with one correction) from Bowie (1879), all rounded off to one decimal place.

protein per day, the subject was in negative balance; at 120 g or above, he was in positive balance. It is unfortunate that no diet provided digestible protein in the critical range from 65 to 105 g. However, even though there is one aberrant negative balance from 133.5 g digestible protein, the results roughly fit a regression line, going through "zero balance" at an intake level corresponding roughly to Voit's standard. The one negative balance in the high-protein periods was explained by a low calorie intake. At the time, the results did seem to provide clear support for Voit's standard of 118 g total mixed protein, equivalent to about 105 g digestible protein, and for his conclusion that feeding an average worker less would cause his body to "degenerate," that is, to lose protein.[53]

Circulating and Organ Protein

In encouraging these simple conclusions, Voit was disregarding his own findings with dogs in the 1860s, even though he had just discussed them again in his 1881 treatise *The Physiology of General Metabolism and Nutrition*.[54] The first finding was that if he took a dog that had been receiving

53. Voit (1881), p. 525.
54. Ibid., pp. 103–18.

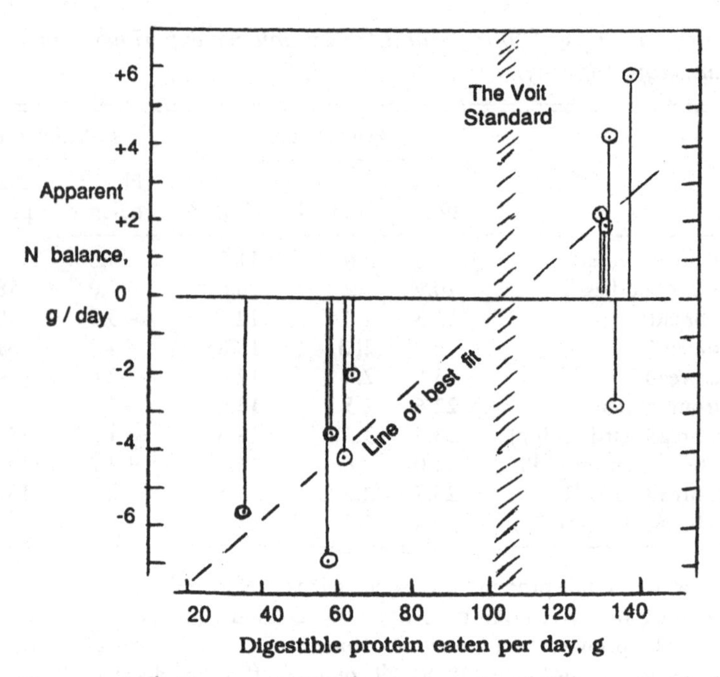

Figure 5.3. Short-term nitrogen balance results of Rubner, as calculated by Bowie and plotted in relation to Voit's standard of 105 g digestible protein for an "average worker of 70–75 kg."

a generous ration of meat for a considerable period and gave it no food for a period of 4 days, its daily output of urinary nitrogen would decline day by day. The average values from six experiments were (day 1) 9.9, (day 2) 3.9, (day 3) 2.6 and (day 4) 1.4 g nitrogen.[55] Dogs of the size that Voit was using would have had at least 700 g nitrogen in their body proteins, so that the fall in excretion rate could not be explained by a significant reduction in total body protein. Voit's explanation was therefore that, in addition to the "fixed" protein that constituted the working muscles and the other body organs, there was a smaller quantity of what he first called *Vorrathseiweiss* – that is, storage or stock protein – and later *circulating protein*.[56] He further hypothesized that organ proteins metabolized slowly, less than 1% per day, and were replaced from the pool of circulating protein. Protein newly acquired from digested food went first into the circulating pool, and this material had a much higher rate of decomposition, perhaps as much as 80% per day, so that it normally provided the majority of urinary nitrogen.

This idea also provided an explanation for Voit's observations that a dog

55. Voit (1867), p. 47.
56. Voit (1869), p. 330.

could be brought into nitrogen balance, within certain limits, at different levels of nitrogen intake. As intake increased, the pool size increased day by day until the daily metabolism also reached a level equivalent to the new intake, and the animal was once more in balance. Similarly, if intake was reduced, metabolism at first exceeded intake; that is, there was a period of negative balance. But this caused the pool of "circulating protein" to shrink, and after a few more days metabolism decreased to a point where it again balanced intake.[57]

Later scholars, considering Voit's model, have drawn the analogy of a wash basin with the drain unplugged and water pouring in from a tap. At one flow rate the basin will remain half full, with the pressure from this height of water pushing water out at the same rate as it is entering, so that the system is in equilibrium. If the flow from the tap is increased a little, the water will at first rise, but the pressure on the exit will increase and eventually the extra flow out will balance the extra in and a new equilibrium will result. Obviously there are limits – with a very low input rate none will accumulate in the basin, and at a very high rate the basin will overflow.

In the light of this model Bowie's calculations look very different. It could simply be said that at nitrogen intakes lower than those provided by the subject's habitual diet (roughly 105 g digestible protein) the subjects were, for short (2- to 3-day) periods, still in negative balance because the new equilibrium had not yet been reached, and conversely for the periods of higher nitrogen intake. On this view the regression line indicated, by the point at which it crossed zero balance, what the subject's habitual intake was, rather than what his requirement was. If he had habitually consumed, say, 130 g digestible protein, then placing him for 2 days on 105 g digestible protein would have put him into negative balance, but would it be reasonable to say that this amount was therefore inadequate?

One cannot help wondering whether Rubner, who had originally obtained the data, presumably under the instructions of Voit as his professor, may have declined to do the calculations of balance, because he had such considerations in mind, so that the job was then given to Bowie. It is known that Voit and Rubner sometimes saw things differently and that the younger man soon moved away to head an independent laboratory at Marburg.[58]

In any case, Voit never seems to have looked more critically at the results of human studies of nitrogen balance. As a medical man he had made out his "prescription," and he seems for the rest of his life to have made use of data that appeared to support it and ignored everything that went against it.

In 1887 a paper was published which demonstrated that gradual adap-

57. Voit (1881), pp. 110–13.
58. Chambers (1952), pp. 4–5; Rothschuh (1975), p. 585.

tation to a reduced protein intake could occur in humans exactly as had been found in dogs. A German physician changed from a high protein diet to one containing only 42 to 45 g protein per day (i.e., approximately 7 g nitrogen), with most of it coming from rice, potatoes and beer and a little from milk and an occasional egg. The values for urinary nitrogen excretion for the first four days were 9.6, 7.5, 6.8 and 5.5 g, respectively.[59] He did not determine fecal nitrogen content, but we can see that a 2- or 3-day balance experiment would have been quite misleading as to the ability of a subject to be in nitrogen balance on such a diet. The author pointed out that in one of Rubner's earlier experiments, in which he gave a 23-year-old, 72-kg soldier a diet consisting of potatoes, butter, oil and salt, and containing 11.5 g nitrogen, the successive urinary nitrogen values for 3 days, were 12.8, 7.6 and 6.0 g. This indicated adaptation and then nitrogen retention on a relatively low protein diet.[60]

Voit and Vegetarianism

Voit had always said that his 118-g standard applied to a mix of animal and vegetable protein and that a higher quantity would be required if vegetable proteins alone were consumed, since they were generally of lower digestibility.[61] This seemed so different from the claims made by vegetarians that he finally took a vegetarian as a subject. He was a 28-year-old paperhanger's assistant, 162 cm tall, weighing 57 kg and described as "well muscled." He had been a vegetarian for 3 years and continued to eat his self-selected food while living in Voit's institute for 14 days. During this period his measured daily intakes were approximately 400 g Graham bread, 125 g pumpernickel, 20 g oil and fruit (apples, dates, figs and oranges). His diet contained 8.2 to 8.9 g nitrogen, equivalent to 53 to 57 g protein (nitrogen content × 6.45), and was estimated to provide 2,550 kcal.

Since the subject did not change his diet, Voit was justified in taking the results for the whole 14-day period. The mean daily intake of nitrogen was 8.5 g and the losses were 8.8 g (3.5 g in feces and 5.3 g in urine), giving a negative balance of 0.3 g.[62] This small value would probably not be regarded as significant now, given the known errors of the analytical methods. A modern physiologist might also predict a small diminution of muscle mass, and therefore a negative balance, because of the inactivity of the subject, in contrast to his usual working life.

Voit was clearly surprised that the subject had done so well and wondered

59. Hirschfeld (1887), p. 547.
60. Ibid., p. 559; Rubner (1879), pp. 147–8.
61. Voit (1876), pp. 21–2.
62. Voit (1889), pp. 255–65.

if this was because his gut had over a long period adapted its function to this diet. He tested how well the same diet would be tolerated by a non-vegetarian. The second subject is described in a heading of Voit's paper as a "worker" (a loaded term to make the contrast with the "vegetarian," who was also, in fact, a working man) and then in the text as "the servant in the Institute, weighing 74 kg, who has been used for most of our balance studies, and no lover of vegetable foods." This appears to have been the same "P.P." who had been used by Rubner and Bowie 10 years earlier and who liked to eat sausages and meat dumplings, taking in something like 120 g protein per day.

He received the same quantities of each food that were eaten by the vegetarian, although he was a bigger man, for a 3-day period. With a daily intake of 8.25 g nitrogen (equivalent to 53 g total protein or 31 g digestible protein) his successive daily excretions of urinary nitrogen showed a declining trend from 11.3 to 9.3 to 8.5 g. Voit took the value of 3.5 g fecal nitrogen per day and averaged the balance over the 3-day period as "−4.95 g N." He added that even taking the third day alone the balance was "−3.75 g N," equivalent to a loss of 24 protein per day from the body. His conclusion was that "the protein in the vegetarian diet did not in the least suffice for the heavier and more vigorous worker; it is essential that he ingests more since he metabolizes 82 g protein per day."[63] Of course, it would be expected that the energy intake was also inadequate and contributed to the negative nitrogen balance.

Analysis of the feces from the two subjects showed no significant differences in the digestibility of the vegetarian diet. In each case the fecal excretion of nitrogen corresponded to just over 40% of the intake. There was therefore no evidence of a long-term adaptation of the digestive system to such a diet.[64]

Voit wrote that the vegetarian metabolized only 39.5 g protein per day. I have not been able to find his basis for this particular value; it may come from urinary nitrogen *plus* a 5-g allowance for protein secreted into the gut and recovered in the feces. Voit accepted that this was an extremely low value, even for a man of only 57 kg, since in proportion to body weight it corresponded to only 51 g protein for someone of 74 kg, the weight of the second subject. His explanation for this finding was that in some way the high starch content of the vegetarian diet had reduced the need to metabolize protein. The vegetarian's mean intake of carbohydrates was 560 g, with sugars accounting for slightly more than starch, so that it was really only high in "total carbohydrates." Analysis of the diet also indicated that it contained 20 g fat, of which 15 g were digested. From these values Voit

63. Ibid., pp. 271–5.
64. Ibid., p. 276.

Table 5.2. *Nitrogen balance in a subject eating different quantities of white bread*

Day	Eaten			N out, urine (feces)[a]	Balance, N (protein)
	White bread	Carbohydrate	N (protein)		
1	698	540	10.5 (68)	12.8 (2.0)	−4.3 (−27.7)
2	834	645	12.5 (82)	12.4 (2.4)	−2.3 (−14.8)
3	1,068	825	16.1 (104)	12.4 (3.0)	+0.7 (+4.5)
Total	2,600	2,010	39.1 (254)	37.6 (7.4)	−5.9 (−38.0)

Note: Data are expressed as grams per day.
[a]Only the total nitrogen content of the feces from the whole experiment was actually analyzed, and the total has been divided into the proportions of bread eaten each day.

calculated that the diet provided altogether 2,550 kcal. In relation to his size this was some 20% higher than the calorie value of the average diet of workers at rest, and the proportion of the calories coming from carbohydrates (88%) was also considerably higher.[65]

He attempted to substantiate this hypothesis, namely, that carbohydrates reduced protein breakdown, by selecting some 1-day values from Rubner's old digestibility experiments. Because Rubner had himself been interested only in the digestibility of a food that had not been thought to vary with the quantity eaten, he had sometimes given different amounts on different days according to how much the subject could consume.[66] On 3 successive days of a trial using white bread, the quantities were increased considerably, and the results were as shown in Table 5.2. As we can see, the subject was responding in the first 2 days in the usual way to a decline in protein intake (i.e., from his habitual 120 g or so), but he was then fed an increased amount and was presumably taking time to react to that. The small positive balance on the third day (which is also the first in which he consumed 104 g protein) may be another transient effect. The carbohydrate content is certainly also increased, but that is not the only variable. Voit seems here to be grabbing at straws in concluding that the improved balance on day 3 was due to the extra starch eaten.

At any rate he sounded confident in his conclusions:

It may be possible for a man to maintain himself in nitrogen balance for some time with a lower intake of protein and a great deal of starch, but such a diet carries

65. Ibid., pp. 264–5, 276.
66. Rubner (1879), pp. 153–4.

disadvantages with it which require one, over the longer term, to return to a diet richer in protein and poorer in starch. It is an interesting fact for physiologists but, from a hygienic point of view, one has to aim always in the direction of using less starch and more digestible vegetables and more protein. . . . The vegetarian exposes himself to disadvantages when he eats a large volume of poorly digested vegetables. It is possible to compose a sufficient and healthy diet just with vegetables for someone with a healthy digestive system, but it is easier with a mixed diet which has a balance of meat.[67]

It is sad to read these unsubstantiated statements from a man who showed such care and precision in his experimental work. Presumably he felt that, as an "authority," he had to sound "authoritative" even when none of the data that he had collected were relevant to the issue at hand. Hindhede, an early critic who analyzed the work in detail, wrote:

Inasmuch as Voit and his pupils did not carry out a single experiment with a moderate worker on less than 118 g protein . . . for any acceptable period of time . . . he could not have had the slightest notion to what extent such a high standard is necessary. . . . When confronted by the question as to what ought to be the daily allowance of protein if a human being is to continue to live in perfect health and vigor, no attention must be paid to Voit.[68]

Hindhede himself had studied several subjects in Denmark who had shown strength and endurance despite living on low-protein, largely vegetarian diets (Figure 5.4).

Nevertheless, Voit's standard, implying the necessity of high protein intakes by vigorous men, continued to be accepted by most textbook writers for the remainder of the century and were also made use of by economists. An Italian writer, on the basis of data indicating that the average worker in Naples in the 1890s consumed only 70 g protein per day, concluded: "It is lack of albumen which renders them so idle, so apathetic, and often so absolutely degenerate."[69] He also quoted from a report which indicated that French laborers had been less productive than the Englishmen working with them in the construction of the Paris–Rouen railway line, until they were given the same portions of roast beef. A few tens of thousands of well-fed English carnivores were, he wrote, able to hold in subjection a hundred million vegetarian Hindus. It was true, he added, that experiments had shown that for short intervals of time people could remain in equilibrium on small quantities of protein, but these experiments were too brief to show the statistical effect of such quantities on the collective life of peoples: "low stature, a low power of resisting disease and slackness in muscular vigor."[70]

67. Voit (1889), p. 288.
68. Hindhede (1913), p. 30.
69. Nitti (1896), p. 62.
70. Ibid., pp. 31, 49–52.

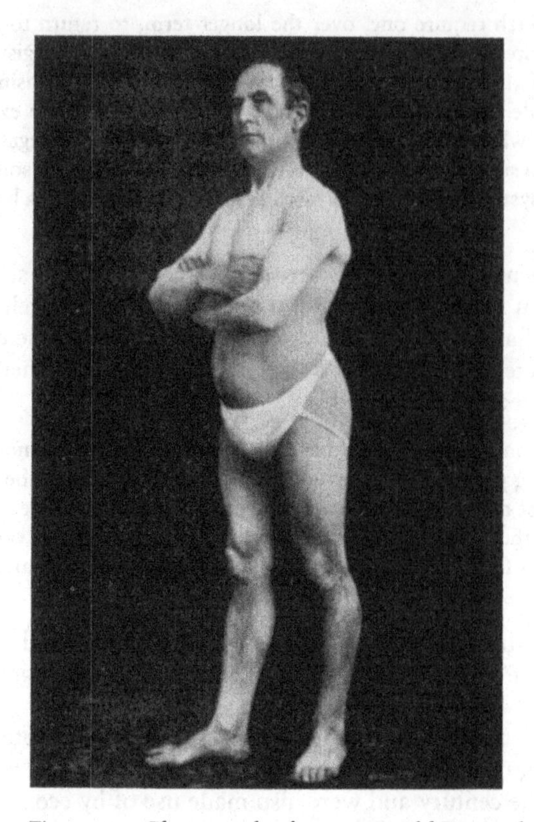

Figure 5.4. Photograph of a 54-year-old Dane who lived on a low-protein diet and cycled 363 miles in 37 hours over boggy roads in a timed trial. (Hindhede, 1913)

The striking claims of vegetarians were therefore disregarded by the orthodox or brushed aside on the grounds either that the subjects were "not real workers" or that the results were "short-term phenomena." But this situation was not to continue much longer, as will be described in the next chapter.

Summing Up

Vegetarianism was originally adopted by some Westerners early in the nineteenth century on the religious or philosophical grounds that it was wrong to kill. U.S. vegetarians later urged that their diet was physically advantageous because it was less stimulating. Kellogg and others argued that it was the high protein content of a meat diet that made it undesirable, both because it was overexciting and because undigested protein in the gut gave toxic breakdown products.

Meanwhile, on the basis of a study of what healthy people unconstrained by poverty actually ate, Voit, in Germany, proposed a standard of 118 g protein per day for an "average worker." A vegetarian subject consuming only 55 g protein per day was found to be in virtual nitrogen balance, but Voit rejected such a diet because it relied on too high a starch content, and his 118 g standard remained generally accepted.

6 Chittenden versus the U.S. Establishment, 1883–1912

As we have seen already, U.S. scholars wishing to gain advanced training in chemistry or nutrition in the nineteenth century typically spent one or two postgraduate (or postdoctoral) years in a German research laboratory. Wilbur Atwater, who has been described as "the founder of the science of nutrition in the United States," was one of these. Born in 1844, the son of a New England minister, he obtained a Ph.D. at Yale in 1869 for a study of the composition of corn (maize) and then spent 2 years at Leipzig and Berlin. In 1874 he became professor of chemistry at Wesleyan University in Connecticut, but had to struggle to find funds for the research in which he was interested. However, in 1879, with the help of grants from the Smithsonian Institution and the U.S. Commission of Fish and Fisheries, he began a systematic study of the composition of North American foods and their economic value as sources of nutrients.[1]

Atwater's Valuations of Foods

In his report to the commission, Atwater included 50 analyses of seafoods (fish, oysters, lobsters, etc.) for water, albuminoids, fats, minerals and extractives.[2] He had followed what was being adopted as the "proximate" system of food analysis – the water (or moisture) value being the loss of weight of the sample after heating at 100°C; the "crude fat" or "ether extract" value being the weight of material extractable with ether from the dried sample; and the "ash" or "mineral" content being the residual weight after the sample had been held in a furnace until the organic matter burned off. The "crude protein" content of the original sample was determined by the Kjeldahl procedure, that is, heating with concentrated sulfuric acid and a catalyst to convert the protein nitrogen to ammonium sulfate, then cooling,

1. Maynard (1962), pp. 3–4; Aronson (1982a), pp. 465–6.
2. Atwater (1883), p. 264.

adding excess alkali, distilling the ammonia and measuring it by titration. Crude protein was estimated as ammonia nitrogen multiplied by 6.25.

Finally, if these various fractions determined for an animal food together came to less than 100%, the shortfall was called "nitrogen-free extractives" (NFE) and considered to represent the small amount of carbohydrate present. For vegetable materials an additional analysis was carried out – the dried, fat-extracted sample was boiled with dilute acid, then with dilute alkali and finally filtered. The insoluble residue was dried and weighed, then kept at red heat in a furnace, cooled and reweighed. The material lost on ashing was referred to as "crude fiber." (Materials like cellulose fall into this category.) NFE was calculated as the shortfall from 100% for the total of all the other analyses including that for crude fiber.

Atwater compared his analyses for fish with German analyses for different types of meat and dairy products. He used a German method for expressing the relative economy of different foods.[3] König had found that in Germany it cost approximately three-fifths as much to buy 1 lb. of fat as 1 lb. of protein, and one-fifth as much to buy a pound of carbohydrate. He therefore prepared an index of the *economic* value of a particular food as

$$\text{protein content} + \tfrac{3}{5} \times \text{fat} + \tfrac{1}{5} \times \text{carbohydrate}$$

He then obtained *relative* values by comparison with average beef, which he used as a standard and gave a value of 100. We can follow the principle with three examples:

Food	Protein (%)	Fat (%)	Extractives (%)	Economic sum	Relative value
Beef	21.4	5.2	—	24.5	100
Halibut	18.2	6.3	—	22.0	90
Oysters	5.8	1.1	3.4	7.1	29

By Atwater's calculations, halibut would have to cost less than 90% of the price of beef before it became an economic choice, and oysters would have to be very much cheaper. In a paper read to the American Association for the Advancement of Science in 1884, he expanded the list of foods to include beans, grains and potatoes and showed that, in these terms, they were more economical than nearly all animal or fish products.[4]

In commenting on his analyses of fish, Atwater had pointed out that true comparisons between fish and meat also depended on their relative digestibility. One suspects that it was the U.S. Fish Commission that gave him funds to carry out such experiments in the winter of 1882–3 at Voit's

3. König (1876).
4. Atwater (1884).

Institute of Physiology in Munich. He used the flesh of haddock, which contained 17% crude protein and only 0.1% crude fat. A test dog first received the fish with a little added fat for 6 days, and then beef steak for another 6 days. In a second experiment a medical student volunteer lived for 3 days on the fish, with a little butter, vinegar and Worcestershire sauce, followed by 3 days on beef, the technique being that described earlier, in Chapter 5. In all four experimental periods the digestibility of the test protein was calculated to be 98%, regardless of whether it was man or dog eating fish or beef.[5]

Probably no one had doubted that fish was a class of well-digested food, so that the direct results of Atwater's work in Munich are not very important. However, Atwater's admiration for Voit's techniques and his adoption of Voit's ways of thinking were to have considerable consequences for the direction of nutrition research in the United States.

Definitions and Standards

Atwater's next project was to collaborate with the Massachusetts Bureau of Labor Statistics in analyzing the findings from a study of the foods consumed by different groups of workers in that state. The final report was obviously written by Atwater, although no author is stated and he is thanked only in the introduction of the volume of the bureau's annual reports for 1886.[6]

He began with a general discussion of the components of foods, and with regard to protein he wrote that it was in general the most important, as well as the most costly, of the nutrients. It served to repair the muscles, tendons, skin and other organs that were being worn out by constant use, but that part was probably transformed into fat and into glycogen, a carbohydrate that occurred in the liver.[7] He added that the dietary standards proposed by Professor Voit were more commonly accepted than any others. They had been estimated in two ways: first, by observing the amounts actually consumed by people whose life circumstances would permit reasonably good nourishment and at the same time preclude any considerable waste of food and, second, by direct experiments.

Atwater gave Voit's standards as 100 g protein per day for an old man, 118 g for a laborer performing a moderate amount of work and 145 g for one doing arduous work. The corresponding total energy standards for these classes were 2,400, 3,000 and 3,300 kcal, respectively.[8] He commented that

5. Atwater (1888).
6. Anonymous (1886), p. 239.
7. Ibid., pp. 243–4.
8. Ibid., p. 262.

these figures represented only general averages, but that if an ordinary laboring man, doing an ordinary amount of work, had much less protein and kept up his muscular exertion, he would be apt, sooner or later, to suffer.[9]

The First U.S. Dietary Surveys

Atwater went on to summarize analyses of the foods purchased by different social groups. Some groups comprised families, and others consisted of workers living in boardinghouses. The values were then expressed as "per man per day" on the assumption that women's needs, on average, were 80% those of men; those of children under the age of 2 were 25%; those between 2 and 6 were 50%; and those between 6 and 15 were 70%. Children over the age of 15 were assumed to have adult requirements.[10]

The averages of results obtained from three boardinghouses were 126 g protein (52% from animal sources), 192 g fat and 539 g carbohydrate per man per day. This we can equate to a total of nearly 4,400 kcal. More controlled studies of food consumption in later years have led us to expect a maximum need of about 3,200 kcal per man. Of course, more physical work was required during that period, but even in a boardinghouse where the occupants were "operatives in a shoe factory, clerks and dressmakers" the daily food purchases were equivalent to 3,900 kcal per man, and in a household where the husband was a skilled printer (i.e., did not perform heavy manual labor) the corresponding value was 4,100 kcal.[11] One can imagine that at a boardinghouse the staff also ate from the general stock and even took home to their families portions that could be regarded as "surplus," but this would not apply to a workingman's family. Atwater was surprised that the fat values were so much larger than those in the records accumulated in Europe: 190 to 200 g versus 40 to 70 g for European men considered to be adequately fed, or 56 to 100 g by Voit's standards. He thought that the overall waste of food was probably higher among the working class in Massachusetts than in Europe, and that this reflected their being better off, but that nevertheless true consumption of food was also higher.

Atwater noted that it was a common practice of the poor to purchase more expensive foods, especially meats, when "food obtainable at only a fraction of the cost would be equally wholesome and nutritious."[12] Here he was referring back to tables included in the preliminary section of his

9. Ibid., pp. 263–5.
10. Ibid., p. 267.
11. Ibid., pp. 290, 306–9.
12. Ibid., pp. 324–6.

paper showing that an expenditure of 25 cents on different animal foods bought very different quantities of protein: beef sirloin, 68 g; round of beef, 132 g; salmon, 18 g; smoked herring, 380 g; and skim-milk cheese, 545 g.[13]

In a series of articles that appeared in the *Century Magazine* over the next two years, Atwater elaborated this theme. He referred to a coal laborer who "made it an article of faith to give his family the best of flour, the finest sugar and very best quality of meat.... [and] spent $156 per year on the meat alone and only $72 for rent in a crowded tenement-house where they slept in rooms without windows or closets."[14] Since food was the principal item of expense for most people, economies here would allow for better housing and the possibility of building a reserve of savings.

Atwater had been hurt by the president of Wesleyan University having told him that his type of work "was not in consonance with the intellectual dignity of a university." He referred to this in his article and replied with an emotion rarely seen in his writing:

> The place of the scholar, as of the saint, was once that of the recluse; now they are both busy among their fellow-men and doing their best to help them.... At the great universities which are the fountain heads of knowledge, speculative philosophy and technology, Sanskrit and sanitation, are studied side-by-side with equal intellect and ardor.... This is following the precept and the example of the great Teacher... a part of whose work on Earth was to feed the hungry and heal the sick.[15]

It seems that Atwater was not the only one the president had offended. By the time Atwater's article appeared he had been "egged" by the students and described in their newspaper as "odious," and on the combined recommendation of the professors and a committee of alumni, the trustees had dismissed him.[16]

Naomi Aronson, a sociologist, has argued that, in fact, Atwater's writing went against the interests of the working class. The readers of his articles were mainly middle-class people with an interest in social problems, and the impression they would have gained was that the poor could help themselves and manage quite well if only they would eat more sensibly. This would, in turn, have reduced their willingness to support workers' demands for higher wages. The hidden implication was that workers had no "entitlement" to the pleasures of the table, with a variety of fruits and vegetables, but only to the tougher cuts of meat requiring extra hours of cooking by the woman of the family.[17]

13. Ibid., p. 254.
14. Atwater (1887–8), p. 437.
15. Ibid., pp. 445–6.
16. Peterson (1964), pp. 115–18, 228–9.
17. Aronson (1982a), pp. 482–3; Aronson (1982b), p. 52.

It may well have been the case that some large employers thought "nutritional science" desirable because it would show that workers could live more cheaply. Yet I believe that Atwater was sincerely trying to provide poor families with a wider range of alternatives for spending their money and that it never occurred to him that his writing might have a negative effect.

Federal Funding for Nutritional Studies

In 1887 the Hatch Act was passed in the United States, providing for an agricultural experimental station in each state. In the next year Atwater was appointed the coordinator of their work and served in Washington for the next three years. His personal interest was, of course, particularly in nutritional studies, and special federal funding for these began in 1894 with Atwater as coordinator. He remained in Connecticut, but stimulated people in different regions to make dietary surveys, coordinated their work and in many cases published his comments as an appendix to their findings.[18]

In 1891 Atwater explained his belief that U.S. standards should be higher than those developed in Europe. First, his subjects apparently ate more, even after allowing for waste. Second, the European standards were based on people whose "plane of living" was lower that of people in the United States, and "to bring a man up to the desirable level of productive capacity, to enable him to live as a man ought to live, he must be better fed."[19] His new standards, including "125 g for a man at moderate muscular work," were "a compromise between the currently accepted European standards and the actual dietaries observed in New England."[20] Atwater added that in the United States many people were eating well above their needs and might be damaging their health as a result.[21] He admonished the farming community for producing overfat meat and mainly low-protein plant crops. He felt they should be concentrating on the production of lean meat and legumes.[22]

Typical of Atwater's comments on the results of the dietary studies were those he made on results from Purdue University. For three mechanics' families, the average daily intakes *per man unit* were calculated to be 105 g protein and 3,570 kcal. Atwater wrote that "the so-called dietary standards are for the most part based on the observed facts of food consumption ... [and] are to be understood to represent simply tentative estimates ... and

18. Maynard (1962), pp. 4–5.
19. Atwater (1891), p. 148.
20. Ibid., p. 149.
21. Ibid., p. 161.
22. Ibid., pp. 168–71.

in no sense to be considered exact or final... but viewed from this point the dietaries are deficient in protein."[23]

In most studies people's diets were found to be at least in the neighborhood of the standards, which is not surprising since Atwater had adjusted the standards in order to make them closer to the intakes found in his first studies. The great exceptions were the families of black farm laborers in the southern region of the United States. Here the food purchased consisted mainly of fat pork, cornmeal, wheat flour and molasses. In addition, some of the families grew their own greens and owned a cow that yielded milk. For 18 families the average intake (per "man unit," as usual) was 62 g protein and 3,270 kcal, compared with average intakes for 50 families in the poorest parts of Philadelphia and Chicago of 114 g protein and 3,330 kcal. As a northerner, Atwater probably felt it politic to make little comment; he therefore limited himself to stating that more extended surveys were needed and that "while the diet of the negro in the South is a very important fact of his character and condition, its effect is hardly to be separated from that of the other conditions of his existence."[24]

Atwater's biographer has pointed out that he maintained his high protein standards even though he knew that men could remain in nitrogen balance on 44 g protein per day and also that muscular work did not increase nitrogen losses provided that energy intake was adequate.[25] He was probably influenced by a German paper that he reprinted, in translation, in the *Experiment Station Record*. There Professor Zuntz argued that although steady, moderate physical work did not increase the need for dietary protein, intensive work over a short period, which led to a deficiency of oxygen in the tissues, resulted in increased metabolism of protein.[26]

Atwater referred to Zuntz's idea when commenting on the results of dietary studies of university boat crews. Their daily intakes had averaged 155 g protein and 4,085 kcal. He concluded that the uniformly high intake of protein by athletes in training must indicate a real demand, rather than individual idiosyncrasy, and that this seemed to confirm "a view coming to be entertained by many physiologists that metabolism... is regulated not simply by muscular work but also by the nervous effect required for its performance."[27] A more skeptical reviewer might attribute the high protein intakes of athletes, first, to their eating more to meet their higher requirement for energy, and thus necessarily to their consuming more protein also if they maintained a normal dietary pattern, and, second, to their being fed at a special table and encouraged by their coaches to eat meat in the belief that

23. Atwater and Woods (1896).
24. Atwater and Woods (1897), pp. 22, 64–5.
25. Maynard (1962), p. 7.
26. Zuntz (1897), pp. 549–50.
27. Atwater and Bryant (1900), pp. 69–71.

this would develop their muscles. Atwater's observation that this was happening, and his conclusion that it represented demand, would thus encourage coaches to continue the practice in a self-fulfilling circle.

Atwater had other nutritional interests. He and Edward Rosa, the professor of physics at Wesleyan, built and operated the first human calorimeter in the United States, in which heat output could be measured directly, as could the balance of gas exchanges. They used this to measure the mechanical efficiency of human work, including mental activity, and the use of alcohol as an energy source.[28] In 1904 Atwater suffered a stroke and died 3 years later.

Professor Chittenden's Doubts

Russell Chittenden, the next important contributor to our story, was born in 1856. Like Atwater, he grew up in Connecticut. His family was not well off, but he managed to study at Yale University, which was near his home, by working as a technician in the chemistry laboratory and assisting with lecture demonstrations. For his bachelor's degree he carried out a small research project on the composition of scallops, showing that they contained free glycine. This was an unexpected finding, and his paper was accepted for publication in Germany. This, in turn, enabled him to spend a year in a leading laboratory in that country. He was then employed again in teaching at Yale and became the country's first professor of physiological chemistry in 1882, at the early age of 25.[29]

For the next 20 years Chittenden's main, ongoing line of research involved the action of different digestive enzymes on food proteins. He had not been directly concerned with dietary standards for humans. However, at the turn of the century, considerable interest had been aroused by the claim of a wealthy, middle-aged American, Horace Fletcher, that if one chewed every morsel of food eaten until it was in very fine suspension, the quantity of food one required would be much reduced. Fletcher offered to serve as a volunteer subject for research at Yale and spent some months as Chittenden's houseguest in 1902. Fletcher's physical condition was certainly exceptional, and the amount of food, including protein, that he ate was remarkably small. Chittenden thought that the main effect of the prolonged chewing was to reduce appetite rather than to increase digestibility. The observations, however, made it seem doubtful that dietary requirements were really as high as was currently believed, and Chittenden decided to investigate the matter in a systematic way.[30]

28. Maynard (1962), pp. 6–9.
29. Vickery (1945), pp. 59–68.
30. Ibid., pp. 80–1.

In November of the same year he gradually reduced his overall intake of food, particularly that of meat, cutting out breakfast altogether and taking only a light lunch. Over the next 7 months, he lost 7 kg of his original weight of 65 kg, but felt in better health, was free of headaches and "rheumatic trouble in the knee joint" and fully maintained his energy for both mental and physical activity. His weight from then on remained stable.[31]

Chittenden did not refer to any calculations of his protein intake during this period, but it seems that he must have made them. In the next year, when he continued the same type of diet, his published values showed a mean intake of approximately 40 g protein per day. Allowing for his relatively low weight, this was still equivalent to no more than 48 g for the "average man" weighing 150 lb. (68 kg), compared with Atwater's standard of 100 g for even an "aged man."

Human Studies

Chittenden was sufficiently impressed by the importance of his finding to organize quantitative studies starting in the fall of 1903 with three different groups of subjects. In the first group were Chittenden himself, along with four colleagues at Yale, all busy, but at work not requiring muscular strength. The second group consisted of 13 soldiers (corpsmen normally employed in army hospitals). During their 6 months at Yale their daily routine included 2 hours in the gymnasium and another hour of drill. The third group, which started in January 1904 and spent 5 months in the study, comprised seven leading student athletes at Yale who continued to train and compete in their sports.

Every subject's urine was collected daily throughout the trial and analyzed in duplicate for its total nitrogen content by the Kjeldahl procedure. For the first 1 to 2 weeks all the subjects remained on their habitual diet. Then they gradually reduced their protein intake and, to a lesser extent, their total energy intake. Also, typically for two periods of 4 to 7 days each, they made a complete record of the weight of each food item that they ate; their feces were also collected and analyzed for nitrogen. The nitrogen intake was also calculated after analysis of each of the foods eaten, and energy intakes were estimated from Atwater's published tables of values for different foods.

The way in which Chittenden handled the data and drew conclusions can be illustrated for a single subject, 32-year-old Lafayette Mendel, whose later work will be discussed in the following chapter. At the beginning of the study he weighed 76 kg; in the following 3½ months his weight fell gradually to 70 kg and then remained steady for the remaining 4 months.[32]

31. Chittenden (1904), pp. 19–22.
32. Ibid., pp. 52–76.

Table 6.1. *Food eaten by Mendel in one day of Chittenden's study and its calculated nitrogen content*

Food	Weight of food eaten (g)			N (%)	N (g)
Bread	58 +	120 +	67.5	1.71	4.20
Coffee (plus milk)	210 +	107 +	210	0.11	0.58
Sugar	21 +	21 +	21	—	—
Custard	—	76	—	0.83	0.63
Milk	—	250	—	0.48	1.20
Potato	—	—	150	0.37	0.56
Lima beans	—	—	80	0.90	0.72
Apple dumpling	—	—	131	0.72	0.94
Candy	—	—	27	—	—
Total					8.83

In the middle of October, Mendel began to modify his diet, mainly by reducing and then eliminating meat, but urine collection and analysis started only on October 26. The average daily urinary nitrogen content was about 12 g during the first week of collection, equivalent to the metabolism of about 75 g protein. It continued to fall for another week, then from November 10 to the end of the study, on June 23, the mean daily excretion was 6.53 g, which Chittenden considered equivalent to the metabolism of only 41 g protein per day.

The following daily average results were obtained during the two balance periods:

	Period I (Feb. 9–14)	Period II (May 18–24)
Energy intake (kcal)	1,975	2,448
Ingested N (g)	7.83	9.19
Urinary N (g)	−7.51	−6.31
Fecal N (g)	−1.48	−1.50
Balance (g)	−1.19	+0.38

The diet varied considerably from day to day; estimated energy intakes varied from 1,700 to 2,300 kcal in the first period and from 1,900 to 3,200 in the second. Daily intakes of nitrogen ranged from 6.5 to 10.4 g. A typical day's food record in the first period for Mendel's three daily meals was calculated to contribute 1,930 kcal and consisted of the foods shown in Table 6.1.

Chittenden commented that the negative balance in the first period was probably due to inadequate energy intake, which resulted in an increase in tissue protein metabolism. He added that having to weigh and record food intake led unconsciously to a reduction in the quantity and variety of food eaten. He pointed out that Mendel had lost 1.2 kg in weight during this 6-day period, and certainly the records for the weeks preceding and following the balance period show lower mean values of 6.82 and 6.45 g, respectively, for daily urinary nitrogen.

He concluded that Mendel, who was a big man and on his feet all day working in the analytical laboratory, clearly needed more than 2,000 kcal/day to be in energy balance. Therefore, the results from the second balance period, when his energy intake was greater and he kept his weight constant, were more representative. He further concluded that with Mendel being in nitrogen balance when he had a urinary excretion rate of 6.31 g nitrogen per day, it was safe to conclude that his average daily metabolism, over a period of 7 months, of 41 g protein (i.e., 6.53 g nitrogen × 6.25) was fully adequate to maintain his health and strength. Allowing for the loss to the feces of 18.3% of the ingested nitrogen, the actual total intake needed was 50 g, which corresponded to 0.715 g protein per kilogram body weight.

The reader will notice that no allowance has been made for daily loss of protein from the body by way of hair and nail growth, sweat or rubbing off of skin. These miscellaneous losses are thought (as we shall see in Chapter 11) to be no more in normal circumstances than 0.03 g protein per kilogram body weight. After allowing for partial indigestibility as before, this would raise Mendel's estimated requirement slightly to 53 g, or 0.753 g/kg. But this is still less than one-half of Atwater's standard.

The records for the other four subjects in the first group indicate that they all remained healthy and vigorous on estimated protein intakes (adjusted to the standard 68 kg body wt) of 46, 58, 74 and 70 g/day, respectively. We do not, of course, know whether they would have remained as fit on even lower intakes.

Soldiers and Athletes

Eleven soldiers completed the study.[33] Two dropped out, apparently finding it intolerable that they were not allowed to go out on the town in their free time. The restriction was to prevent unrecorded consumption of food and drink. Over the study, the subjects' average weight fell very slightly, from 62.1 to 61.4 kg, and all the men showed an improvement in physical condition as assessed by their medical officer, and from the results of strength tests (Figure 6.1). The level of hemoglobin in their blood was also fully maintained.

The soldiers' urinary excretion values for the first 2 weeks, when they

33. Ibid., pp. 131–326.

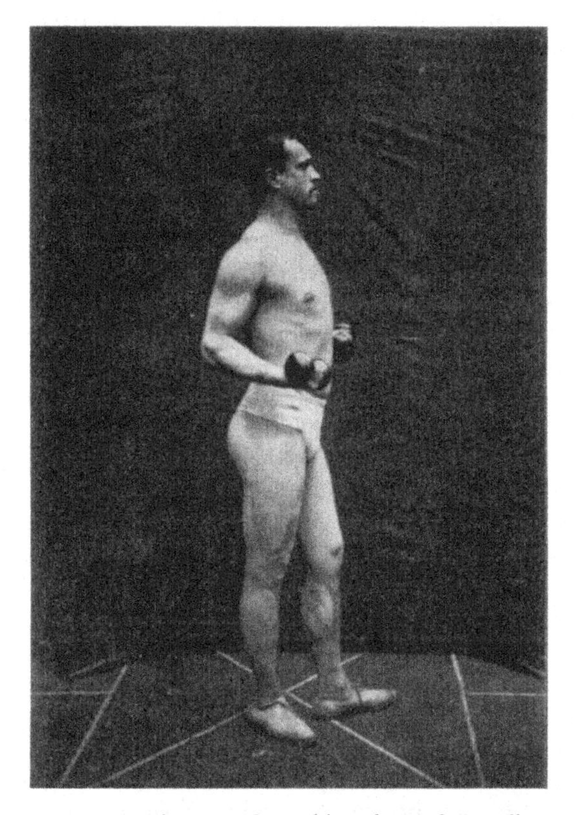

Figure 6.1. Photograph used by Chittenden to illustrate the good physical condition of his subjects after 5 months on a low-protein diet. (Chittenden, 1904)

were still on regular army rations, indicated an average protein consumption of about 105 g/day. Once the diet pattern was fully changed, the daily intake fell to about 55 g/day (which corresponded to 61 g for the standard 68-kg man) and the individuals' average urinary nitrogen excretion per kilogram of body weight ranged from 0.106 to 0.150 g. This was similar to the range observed in the first group.

In the first balance period, the men were all in negative balance, but the energy intake was only about 2,000 kcal/day. By the second study period, intakes had increased to 2,500 kcal and there was an average positive balance of 0.6 g nitrogen. The one man in negative balance was described in the gymnastic report as being "nervous, irritable and aggressive." If an uncooperative subject were to scrounge extra food from the kitchen, this would of course increase his urinary nitrogen values and make him appear to be in negative balance in relation to his recorded intake. This is at least a possible explanation, since this subject did not lose weight over the trial and gained in strength.

The third group, consisting of student athletes, were in the study for only 5 months (from mid-January to mid-June).[34] In the initial week, while still on their usual diet, their urinary nitrogen values indicated they were taking in, on average, about 130 g protein per day. By the last 2 months the level of urinary excretion had almost halved. On average the men lost 2.7 kg in body weight during the trial, mostly in the first 2 months. However, they said that they felt better, their strength tests improved and their athletic performances were at least as good as previously.

Only one balance study was carried out with this group. From the results, two of the seven subjects appeared to be in negative balance. Chittenden again attributed this to these two having reduced their intake as a consequence of the measuring required. Certainly this effect was seen in the first balance periods of the other groups, for whom there was more than one balance study. From the data as a whole, it appeared that 64 g protein would be adequate for an average 68-kg man engaged in athletics. The corresponding Atwater standard would again be roughly double this value.

We see, therefore, that the three groups gave very similar results, and Chittenden felt justified in concluding that people could safely live on much less protein than the standards set out by Voit and Atwater. Further, they did not require excessive quantities of carbohydrates in order to achieve nitrogen balance, as had been argued by Voit in his paper on vegetarianism that was discussed in the preceding chapter. Chittenden's second conclusion was that low-protein diets were actually more healthy than high-protein diets. He believed that this was because the body had smaller quantities of uric acid and other protein breakdown products to dispose of, which explained the general feeling of well-being experienced by the subjects in the trials.[35]

The Great Debate

Chittenden's publication made a considerable stir, even though his conclusions were not entirely original. As we saw in the preceding chapter, vegetarian groups, including the Seventh Day Adventists, had long claimed that their low-protein diets kept them in superior health. But orthodox scientists and physicians could, and did, shrug off such claims on the grounds that they came from enthusiasts who reported mainly their feelings and were therefore unscientific. Also, although several scientists had reported exper-

34. Ibid., pp. 327–453.
35. Ibid., pp. 454–63.

iments on themselves in which they remained in nitrogen balance on low-protein diets for periods of 1 to 2 weeks, it was easy to consider this a purely short-term phenomenon.

The work at Yale could not be dismissed or ignored so easily. Professor Chittenden was himself a member of the scientific establishment and had presented his results to the National Academy of Sciences. He had also worked with physically active men, as well as relatively sedentary academics. In addition, he had continued the studies for many months and accumulated quantitative data on strength and physical performance, as well as on nitrogen balance.

By 1906 Francis Benedict had taken over Atwater's program, and in a paper published in that year he paid tribute to the quality of Chittenden's work. It was "scientific research of the highest type, and for the most part defying criticism."[36] However, he was still unwilling to accept his conclusions. With regard to the soldiers, he felt that their steady routine with regular moderate exercise and no opportunity for dissipation would have led to an improvement in physical condition even if their diet had been unchanged. No individual was worked to a condition of fatigue, when the wear on tissues was out of proportion to the work performed. With regard to the athletes, he wrote that it was difficult to understand why, if the low-protein diet had really suited them so well and helped their performance, they had almost all (so he had heard) reverted at the end of the study to a more traditional diet rich in meat.[37]

As he and Atwater had written earlier, "A man may maintain bodily equilibrium on either a higher or lower nitrogen level." The answer as to which was most advantageous "must be sought in broader observations... efficiency, health, strength and welfare."[38] He, Benedict, remained more impressed by the observation in so many countries that men engaged in heavy work consumed considerably more protein than those engaged in less physically demanding activities:

> Dietary studies made in England, France, Italy, and Russia show that a moderately liberal quantity of protein is demanded by communities occupying leading positions in the world.... It certainly seems more than a remarkable coincidence that peoples varying so widely in regard to nationality, climate and geographical conditions... should show such agreement.[39] The negro and poor white of the South, the laborers in Southern Italy, all partake of diets relatively low in protein. That their sociological condition and commercial enterprise are on a par with their diet, no one doubts.

36. Benedict (1906), p. 424.
37. Ibid., pp. 418–19, 421–2.
38. Atwater and Benedict (1902), p. 130.
39. Benedict (1906), p. 416.

...when people accustomed to a low protein diet are fed on a higher protein plane, as is the case when Southern Italians come to America, their productive power increases markedly.[40]

Another writer expanded on this point:

The only people who fail to attain the ordinary [Voit] protein standard are those who are too poor to afford the cost.... As soon as financial considerations are surmounted, the so called 'vegetarian Japanese' or Hindu raises his protein intake to reach the ordinary standard of mankind in general.... If Chittenden is right, then all the world up to this time, with the exception of a few faddists, has been wrong. It is 'Chittenden against the world' and it is inconceivable that all mankind, under the most diverse conditions, should have fallen into the same mistake.[41]

Criticism from India

The most systematic attack came from D. McCay, a British professor of physiology in Calcutta. He had been greatly impressed by the differences in physique among members of the various communities in India. In particular, the typical Sikh in the Punjab was muscular and vigorous in temperament, whereas the Hindu Bengali was usually, in his experience, relatively flabby, lethargic and incapable of performing a hard day's work. He related these differences to the characteristic diets of the two groups. He found that the Sikh community, with wheat as its staple grain, consumed generous quantities of meat and milk, so that an adult male could receive as much as 135 g protein per day. By contrast, the average Bengali's rice diet was supplemented with only small quantities of animal protein, and men typically received only 54 to 65 g protein per day. By Chittenden's standards this intake should have been healthful, and indeed optimal, but in fact the Bengalis were not virile and energetic.[42]

McCay accepted that men could remain in nitrogen balance on low levels of protein. Bengalis, for example, must do so because they spent a lifetime on such diets. But whereas these low amounts were adequate for existence, the question remained as to what quantity was required for maximum efficiency. McCay reprinted a quotation: "Energy is not to be confused with muscular strength... [it] is a property of the nervous system.... Muscles give us the power to work; the nervous system gives us the initiative to start it.... If protein be regarded as a nervous food... it is not without reason that the more energetic races of the world have been meat-eaters."[43]

McCay, too, praised Chittenden for his work and for doing so much to

40. Ibid., pp. 427–8.
41. McCay (1912), p. 102.
42. Ibid., pp. 78–9, 90–2, 153–6, 203–4.
43. Ibid., pp. 106–7, 153.

"awaken interest and lift the subject of dietetics from the playground of quacks and charlatans."[44] But he went on to subject Chittenden's results to detailed examination. He first criticized the use of urinary nitrogen as a direct measure of protein metabolized. According to further results published by Chittenden in 1911, six men working in his laboratory had changed to a low-meat diet; both their food intakes and total excreta had been recorded and analyzed continuously for 18 weeks.[45] All gave similar results, the mean values (expressed as grams nitrogen per day) being as follows:

Intake	12.02
Fecal loss	−1.36
Urine loss	−9.16
Apparent balance	+1.50

McCay said that these results demonstrated that urinary nitrogen gave an underestimate of protein metabolized. Since there had been a net absorption of 10.66 g nitrogen (i.e., 12.02 − 1.36 g), and the men were not said to have been growing or gaining weight, this quantity must have been metabolized – some through the skin, and some lost as hair and so on. The difference of 1.5 g "lost" nitrogen (equivalent to 9.4 g protein) should therefore be added to all of Chittenden's estimates of metabolized protein in his main experiments, where food and fecal nitrogen values were not available for the full periods of the trials.[46]

If we apply this correction to the results that Chittenden gave for Mendel's experiment, which were examined in detail earlier, 1.5 g must be added to the 6.53 g nitrogen found in the urine. This makes 8.03 g nitrogen (equivalent to 50.2 g digested protein). Allowing for incomplete digestibility as before, we find that the revised estimate of Mendel's protein intake when he was in balance is 61.4 g. This is a significant increase, but of course still well below Voit's standards.

Were the Data Reliable?

By the time of McCay's writing, Chittenden himself had remained on his modified diet for 7 years and claimed that he had never been fitter or more mentally active. However, McCay questioned whether the data given for Chittenden's two 6-day balance periods – the only times that actual intakes were determined – were representative or even correct. The calculated daily intake of 1,600 kcal corresponded to only 28 kcal/kg body wt. "Yet the average amount hitherto considered absolutely necessary is about 40 kcal/kg. Either this is far above the true requirements *or* there is some fallacy in

44. Ibid., pp. 68, 145.
45. Chittenden (1911), pp. 659–60.
46. McCay (1912), p. 84.

Chittenden's data *or,* if the results from the experiment are to be extended to mankind in general, the law of conservation of energy in human beings must be seriously called in question."[47] McCay had a strong point. Energy requirements can be determined either directly from heat output or indirectly from respired carbon dioxide, and modern recalculations confirm that someone of Chittenden's weight and age range would be expected to use approximately 40 kcal/kg daily even with very light activity.[48]

With reference to the results obtained with the soldiers, McCay drew attention to the considerable (i.e., twofold) differences in fecal nitrogen content among the subjects when they were supposedly eating virtually the same diet during the balance periods. Thus the apparent percent digestibilities of the nitrogen varied from 75% to 90%. McCay suggested that this was inconsistent with the individuals truly eating the same diet.[49] Benedict had also noticed the variability and considered it to be evidence of malabsorption among some of the men, their diets being too low in protein to sustain optimal body function.[50] However, this is not a satisfactory explanation, since it is the higher values for apparent digestibility that seem out of line.

Examination of the detailed data published by Chittenden indicates another explanation. Considering the full 6 days of the first balance period, which included 12 men, and the first 6 days of the second (7-day) trial, which included 11 men, if every man had produced a fecal sample every day, there would have been altogether 23 fecal samples on the first days of the two trials, 23 on the second and so on. If some men did not have a bowel movement on a particular day, there would be some number fewer each day, but one would expect the number of "missing samples" to be randomly distributed. In fact, in the 6 successive days, the missing totals were 15, 12, 4, 4, 1 and 1. Such a skew distribution could have occurred by chance less than 1 in 10,000 times. One seems forced to conclude that many of the men were unwilling to turn in their fecal material for weighing and analysis in the first portion of each trial. And, indeed, the calculated protein digestibility value for those subjects who said they had no bowel movement on 3 or more of the 6 days is 88%, and that for the remainder of the subjects is 81%. This seems to confirm the idea that not all the fecal material was collected from the first group.

If this conclusion is correct, it certainly indicates that the control of the "soldier experiment" was imperfect, and McCay may have been right also in doubting whether the full amount of the daily urine was always handed

47. Ibid., p. 115.
48. FAO/WHO/UNU (1985), pp. 72–6.
49. McCay (1912), pp. 125–6.
50. Benedict (1906), pp. 420–1.

over for analysis. He quoted another author who had written that some of the soldiers had later admitted that they had eaten unauthorized meals during the experiment.[51] Certainly the soldiers' computed mean energy intake of 2,300 kcal/day does, as McCay points out, seem extremely low in relation to the directly determined energy expenditures in rest and activities of various kinds that have since been recorded.[52]

Chittenden's Response

The British Medical Association held two debates during this period on nutritional requirements, and summaries of the discussions were published. The first was held in Canada in 1906 and the second in England in 1911.[53] Chittenden was the opening speaker on each occasion, and in the second debate he made some new points. One was that critics had tried to argue, on the basis of grams protein consumed per kilogram body weight, that the suckling infant's intake of breast milk, "nature's food," corresponded roughly to the Voit standard for adults. However, as Rubner had recently set out in great detail, human breast milk (in contrast to the milk of other species) was quite low in protein, which contributed only some 10% of its total calories. Further, it was agreed that growth itself required a high proportion of protein, and Rubner had calculated that, of the nutrients remaining for maintenance in the growing infant, only 5% of the energy came from protein.[54] This corresponds to a ratio of 37.5 g protein to 3,000 kcal.

Chittenden also argued that it was the relative prosperity of Westerners that allowed them to consume a high-meat and therefore a high-protein diet. His critics were trying to reverse the true sequence of cause and effect. Moreover, it was more likely that the bad effects of the poorest Indian diets were the consequence of their lack of trace nutrients: "There are many factors, aside from nitrogen and calories, that play a part in determining proper nutritive conditions. I have seen dogs, for example, go to pieces on a vegetable diet, while the substitution of a little meat or milk, with the same total of available nitrogen, is followed by a satisfactory condition of health."[55]

Chittenden was referring here to feeding experiments that he had undertaken in response to reports from Europe that, although dogs remained in nitrogen balance for some time on low-protein diets, they eventually weakened and succumbed. This was a powerful argument for caution in adopting

51. Loeser (1912), p. 156.
52. McCay (1912), pp. 145–8.
53. Anonymous (1906); Anonymous (1911).
54. Rubner (1908), pp. 96–7, 108–9; Chittenden (1911), p. 662.
55. Chittenden (1911), p. 661.

Chittenden's ideas, and he was in turn responding with a bold prediction that the deficiency was one of hitherto unknown nutrients rather than of protein. But to follow this line of investigation and its successive revelations will require a new chapter.

Summing Up

Wilbur Atwater, who was the leader of the first U.S. dietary surveys in the 1880s and 1890s, set the protein standard for physically active men at 125 g/day. This was made slightly higher than Voit's standard on the ground that Americans were more vigorous and lived at a higher level of activity than Europeans. Working people, however, were urged to economize on food purchases by buying cheaper cuts of meat that were just as nutritious as more expensive ones, even if less appetizing.

Russell Chittenden, an established physiological chemist, doubted the validity of this high standard and studied the physical performance and nitrogen balance of volunteers who consumed only about one-half the standard quantity of protein for 6 months, but remained vigorous and apparently in nitrogen balance. However, others were not persuaded that such a diet could be safely maintained for longer periods, because it seemed a general rule in the world that those who could afford to do so ate more and that the poor who did not were generally lacking in energy and initiative. Chittenden tried unsuccessfully to argue that the deficiencies of the "poor" diets were more likely to be those of unidentified trace nutrients than those of protein, and that it was affluence that *caused* people to eat a "rich" diet, and not vice versa, as his critics claimed.

7 Vitamins and Amino Acids, 1910–1950

ONE OF THE SERIOUS CRITICISMS of Chittenden's advocacy of low-protein diets was that, while dogs on such regimes remained healthy for many weeks, they eventually became ill, lost their appetite and died. This had been the experience in two German laboratories. In each the daily diets contained about 1.5 g protein and 90 to 100 kcal/kg body wt and consisted of minced meat (or dehydrated meat powder), animal fat and rice. In one laboratory, where meat powder was used, the dogs' digestion became less efficient; they vomited and lost their appetites.[1] In the other laboratory there was no change in digestibility but, after 6 weeks or so, the dogs lost their appetite and finally died.[2] In both laboratories, autopsies showed fatty degeneration of the gut wall and of the liver.

In the next decade a worker in Finland undertook further trials with two dogs using a simple mix of meat and sugar. The dogs did quite well for many months and digestibility remained normal. But finally each dog died after a short acute illness, which was diagnosed as resulting from a severe infection.[3]

Chittenden, skeptical as ever, wrote:

> We must first be sure of our facts before arguments or conclusions of any kind are warranted. It is to be remembered that dogs are as sensitive in many ways as man, and no physiological experiment covering a long period of time can be carried out with any hope of success unless there is due regard for proper hygienic conditions, some degree of variety in diet, and reasonable opportunities for fresh air and occasional exercise. Is it strange that dogs confined in cages barely large enough to permit their turning around, and fed day after day and month after month with exactly the same amount of desiccated meat, fat, and rice, should show signs and symptoms, if nothing worse, of disturbed nutrition?[4]

1. Munk (1893).
2. Rosenheim (1893).
3. Jägerroos (1902), pp. 389–412.
4. Chittenden (1907), p. 242.

He therefore went on to carry out a large series of experiments of his own.

Chittenden's technique was to alternate the conditions of the dogs between 10 days of confinement in a cage, where urine and feces could be collected in order to determine nitrogen balance, and 3 weeks of running free in an airy room, and even in an outdoor paddock for several hours on sunny days. The diet was also varied from month to month. For example, in one trial it started with meat, lard and cracker dust. Then the cracker dust was replaced by bread, and in the next month half the meat was replaced by milk. The total nitrogen and energy values of the diet were, of course, kept constant. Under these conditions the dogs remained in good health and vigor for many months, though they received only 1.6 to 1.7 g protein and 70 kcal/kg.[5] This contrasted with the conclusion from the German studies that even 1.9 g protein per kilogram was inadequate, and Chittenden attributed their failures to the fact that the dogs were kept continuously in small cages.

However, when Chittenden removed meat and milk entirely from the diets, and the same nitrogen intakes came from bread, beans and/or peas, the animals quickly developed gastrointestinal disturbances and lost their appetite and energy. He wrote that these results could have been due to some specific differences between animal and vegetable proteins or to "the possible need of the animal's body for extraneous principles which only meat, milk, or other animal products can supply.... Other substances without any appreciable fuel value are quite likely to be of primary importance in controlling and regulating the various processes of the body, which combine to maintain the condition of normal nutrition. With a diet restricted to one or two vegetable products, it is quite conceivable that something may be lacking which the system demands, though it cannot be measured in terms of nitrogen or calories."[6]

The Search for Unknown Nutrients

Both possibilities (i.e., the importance of so far unknown nutrients and nutritional differences among proteins) that Chittenden put forward in 1907 were to become subjects of active research. We will begin with the first type of investigation, but in abbreviated form since it is peripheral to the primary subject of this book.

Writing in 1905 about the dangers of low-protein diets, Benedict argued that the disease beriberi had been eliminated from the Japanese navy by increasing the protein content of rations to European standards.[7] This had

5. Ibid., pp. 244–52.
6. Ibid., pp. 253–6.
7. Benedict (1906), p. 429.

been achieved by Takaki, who replaced a proportion of the predominantly "white rice" diet with wheat bread, peas, beef, fish and *miso* (fermented soybean).[8]

Beriberi was sometimes referred to as the "national disease of Japan" because of its prevalence there, but it was a problem in other parts of Asia too. It was also to be experienced by many U.S. and British soldiers held in Japanese prison camps in World War II. The principal symptoms were weakening of the heart and degeneration of nerve fibers, particularly in the legs, causing paralysis. The mortality could be very high.[9]

By the 1880s European doctors were seeing beriberi in British Malaya and the Dutch East Indies. And with the great influence of the "germ theory of disease" at that time the disease was generally thought to be either a direct infection or the result of poisoning from a toxin produced by a microorganism. Research was stimulated by the observation that a disease in chickens characterized by damage to peripheral nerves was associated with their being fed white rice. The disease did not occur if their diet was supplemented with "rice polishings," the fragments removed when brown rice was refined to white rice by abrasive milling. Further inquiries showed a comparable distribution of beriberi among prisons in both countries; that is, it was a continuing problem where white rice was the prisoners' staple food, but not where they were given brown rice. The hypothesis first advanced to explain this finding was that white rice contained a toxin, but that a benevolent Nature had provided the antidote in the outer layer of the grain.[10]

There was little difference in the protein content of the two forms of rice. Moreover, the "chicken disease" could be produced with a number of starchy foods, so that it was not due to a specific effect of white rice. It could also be cured by microscopic quantities of extracts made from either rice polishings or brewer's yeast. In one series of tests, even 4 milligrams (mg), equivalent to 1/7,000 of an ounce, of a still impure extract was a curative dose.[11]

The Discovery of Vitamins

In 1912 Casimir Funk, a 28-year-old Polish chemist attempting to isolate the factor in rice polishings at London's Lister Institute, put forward the bold hypothesis that there were at least four organic compounds that must be present in our diet (though at quite low levels) in order to prevent the

8. Takaki (1885); Anonymous (1887a).
9. Williams (1961), pp. 53–8, 68–80.
10. Ibid., pp. 36–48.
11. Funk (1913), p. 179.

following four diseases: beriberi, scurvy, pellagra and rickets. He further hypothesized that each was a compound containing an amine ($-NH_2$) group, and he named these hypothetical vital factors *vitamines*.[12] For the next 30 to 35 years a large proportion of the nutritional research carried out throughout the world was devoted to establishing the identity of these new organic nutrients. It soon became clear that they were not all "amines," so the term was shortened to *vitamins*. The general history of these complex developments has already been well covered,[13] and I have myself contributed volumes on pellagra and scurvy specifically.[14]

An early finding was that young rats failed to grow normally when given "purified" diets consisting of protein, fat and carbohydrates together with a mixture of minerals. However, normal growth rates could be obtained if these diets were supplemented with butterfat (or cod-liver oil) and skim milk from which the protein had been removed. It was concluded that butterfat and cod-liver oil contributed an unknown fat-soluble factor "A," and the protein-free milk, a water-soluble factor "B." After a short time they were named "vitamin A" and "vitamin B."[15] It was then realized that the active factor in lemon juice and cabbage that cured an experimental scurvy-like condition in guinea pigs was neither of these, and it was named "vitamin C."

After that, things became more complicated. More fat-soluble factors were discovered and given the letters D, E and K. It was also realized that vitamin B was a complex, and individual factors preventing different conditions were coded B_1, B_2 and so on. However, some of the discoveries remained unconfirmed, and the system became muddled. Therefore, as soon as the water-soluble vitamins were isolated and identified, they were generally given names reflecting their chemistry. However, cobalamin (the cobalt-containing vitamin) is still remembered by its original code of B_{12} and pyridoxal (containing a pyridine ring) is often referred to as B_6, partly because several slight variations of pyridoxal have the same type of biological activity.

By 1950, as a result of collaborations between nutritionists working with laboratory animals and organic chemists, at least 14 vitamins had been isolated and identified. In addition to the diseases listed by Funk, blindness among Third World children was often due to a deficiency of the first fat-soluble factor (vitamin A). In the same period it was also realized, or confirmed, that people could suffer from deficiencies of inorganic elements that were needed only in very small quantities. For example, deficiency of iodine

12. Funk (1912).
13. McCollum (1957), pp. 201–318; Guggenheim (1981), pp. 111–276.
14. Carpenter (1981); Carpenter (1986a).
15. McCollum (1957), pp. 214, 291.

resulted in goiter and cretinism, and that of zinc in dwarfism and dermatitis. Iodine and zinc are therefore two of the "trace elements" required in our diet.

The Japanese succeeded in eradicating beriberi from their navy by giving sailors a greater variety of foods that were richer in protein than white rice was. However, it is now appreciated that these additional foods were also richer in thiamine, the antiberiberi vitamin, and that it was this, rather than the additional protein, which explained the success.

Chittenden's provisional hypothesis, that his dogs maintained on a diet of dried peas, cracker meal and lard died because of a deficiency of an unknown trace nutrient, also proved to be correct. It was found that the disease condition could be prevented by supplementing the dogs' diet with a small quantity of carrots, because, as was later shown, carrots contained the yellow pigment carotene, which animals can convert to the active form of vitamin A.[16] The observation that animals could do well for a period on a diet lacking vitamin A was explained by their accumulation of a reserve of the vitamin in the liver during an earlier period on a rich diet consisting, for example, of dam's milk, and the gradual use of these reserves. In general, the water-soluble vitamins are not stored as well in animals and humans, though cobalamin (B_{12}) is an exception, and people may subsist for several years on a vegan diet containing none of this vitamin before irreversible nerve damage occurs. (A "vegan" is a strict vegetarian who takes no milk or eggs.)

Because of the skills of chemists and the very low levels of vitamins that are needed, vitamin supplements can be purchased for only a small proportion of the cost of one's ordinary food needs. In the wealthier countries of the world most people can afford, and in fact choose, a varied diet that meets their vitamin requirements. But in the poorer countries, where many families can afford only a much more limited diet, young growing children are more likely to be vitamin-deficient.

The main problem for governments or other organizations that want to provide aid is the cost not of purchasing vitamins, but of distributing them to those actually in need, often in remote villages. Where staple foods, such as white rice, are processed in relatively large amounts, it is practicable to have the foods fortified with the important vitamin(s) lost during processing. In the use of white rice, a supplement of only 5 parts per million of thiamine is sufficient to prevent the occurrence of beriberi. In most developed countries, white wheat flour is fortified with three or more vitamins to the levels found in whole wheat grains.

Although, as just mentioned, the common "affluent" diets contain sufficient amounts of each vitamin to eliminate any risk of deficiency conditions

16. Underhill and Mendel (1928), pp. 632–3.

developing, some people recommend taking "megavitamin" dosages in order to induce a state of "supernutrition." For example, the ordinary daily recommended intake of vitamin C is 60 mg, but Linus Pauling has written that he takes some *200 times* this quantity, that is, about 12 g/day, in order to reduce his risk of developing cancer and other diseases. The benefits from such a practice are still a subject of controversy.[17] What is noncontroversial is that our knowledge of vitamins and trace elements has provided a better understanding of the effects of food processing and increased the public's ability to make healthy food choices.

The Constituents of Proteins

Once it had been found possible to achieve good growth in young rats with purified diets supplemented with protein-free sources of vitamins and minerals, the way was open for rapid development in understanding what made some proteins apparently of higher nutritional value than others.

We saw in Chapter 2 that gelatin, though rich in nitrogen, could not replace the nitrogenous compounds of meat in the diet of dogs. Then Mulder, in 1839, hypothesized the existence of a chemical unit, called "protein," which, in combination with varied proportions of sulfur and phosphorus, formed albumin, fibrin and gluten (as explained in Chapter 3). But this idea soon collapsed. Although the proteins of vegetable foods had properties and carbon–nitrogen ratios that were very similar to those in animal tissues, they were *not* identical. Nevertheless, the animal kingdom apparently had to use vegetable proteins as a basis for forming their tissues.

Chemists like Liebig realized that they were dealing with large molecules and seemed to visualize that the changes animals made to the albuminoids they ate involved the removal where necessary of a few atoms or radicals and the addition of others, but that when an animal or human ate meat, the proteins could be deposited with virtually no modification. They knew that, by boiling proteins with either strong sulfuric acid or strong alkali, some much simpler crystalline products could be obtained. These – named, for example glycine, leucine and tyrosine – all belonged to the same class, called *amino-bodies* at first and then *amino acids*.[18] However, at the time, this seemed to have no biological significance, since the conditions of breakdown were so extreme, in contrast to the moderate temperature and neutral reaction of animal tissues. (The chemical structures of the amino acids are given in Appendix A.)

During the same period (i.e., the 1830s and 1840s) there was an increase in knowledge of the digestion of proteins. Extracts made from the dried

17. Carpenter (1986a), pp. 210–20.
18. Vickery and Schmidt (1931), pp. 187–213.

mucus of stomach linings changed the solubility and coagulability of dried albumin in acid suspension in the same ways as had been observed in the stomachs of animals killed some time after receiving a meal of dried albumin. The active factor in these extracts was named *pepsin*. However, chemical analysis of the digestion products showed no significant change in the carbon–nitrogen ratio. This allowed both Liebig and Dumas to conclude that there had been no real chemical change in the food protein, "only a change in its state of cohesion," so that it could pass more readily through the gut wall.[19]

After the discovery in the 1860s that pancreatic juice secreted into the small intestine exerted a powerful digestive action on proteins, the free amino acids leucine and tyrosine were recovered in traces in intestinal contents (i.e., in vivo) and in quite large quantities from the incubation of proteins with pancreatic juice in glass bottles (in vitro) at body temperature.[20] At first, German scientists considered that this was probably an artifact from the putrefactive action of bacteria present in the gut. It seemed to them that "such a profound decomposition would be a waste of chemical potential energy, and a reunion of the products of such a profound decomposition is highly improbable."[21]

Little attention was therefore paid to these findings by students of nutrition. It was assumed that under natural conditions an animal organism would make only minimal changes to the proteins it consumed before incorporating them into its tissues. Scientists labored for many years to categorize the products of peptic (i.e., stomach) digestion and spoke of different classes of proteoses, peptones and so on, according to the solubilities under different conditions of the fractions they isolated. Presumably they assumed implicitly that they might discover a universal "core" around which all the individual animal proteins could be built up.[22] There was no proof that whole proteins consisted only of pre-formed amino acid units; the combined yield of amino acids recovered, even from acid hydrolysates in this period (i.e., before 1890), corresponded to less than 50% of the original weight of the proteins.[23]

It had also been observed repeatedly that if a large supplement of a single amino acid such as glycine were given to a dog, almost all of the added nitrogen would be recovered in the urine as urea within the next 24 hours.[24]

19. Holmes (1974), pp. 163–76, 181–6; Matthews (1991), pp. 7–19.
20. Lea (1890), pp. 243–4.
21. Bunge (1902), pp. 167–71; Greenstein and Winitz (1961), Vol. 1, pp. 249–63; Holmes (1979), pp. 179–80.
22. Fischer and Abderhalden (1903), p. 83; Vickery and Osborne (1928), pp. 406–7; Fruton (1979), p. 5; Fantini (1983), pp. 9–15.
23. Moore (1898), pp. 401–40; Chibnall (1939), p. 15.
24. Salkowski (1880), pp. 100–7.

This threw further doubt on the nutritional value of free amino acids and led to an ingenious alternative hypothesis: that the cells of the intestinal wall could limit the absorption of proteoses and peptones to the quantity required by the body for growth and repair. It was the quantities remaining *in excess* in the gut that were then further broken down by pancreatic juice to free amino acids. These were rapidly absorbed and metabolized by the liver to urea, which was excreted.[25] Chittenden, for example, wrote in 1895: "We may well consider the formation of these amino-acids in pancreatic proteolysis as a means of quickly ridding the body of any excess of ingested protein food, with the least possible expenditure of energy on the part of the system. This has always seemed to me the probable purpose of the profound changes which the pancreatic ferment is capable of inducing."[26]

The scientists who believed that proteoses and peptones were absorbed by the gut wall were puzzled by the fact that they were unable to find these compounds in the blood of animals that were digesting a protein meal. They proposed that these molecules disappeared because they were immediately recombined by the cells in the gut wall into ordinary protein before being passed into the blood. In one experiment carried out to test this idea, a horse was given large meals of the wheat protein gliadin, which had been found to yield about 36% of glutamic acid on hydrolysis. Blood samples were taken from the horse before and after the digestion of the gliadin, but analysis indicated no rise in the glutamic acid content of the mixed blood proteins from their normal value of about 8%. It did not appear, therefore, that "re-formed" gliadin was transferred to the bloodstream.[27] Moreover, when other workers attempted to study the changes undergone by proteoses and peptones in contact with isolated pieces of small-intestine wall, they found no evidence of protein synthesis. Instead they discovered another enzyme, *erepsin,* which caused them to break down further to free amino acids.[28]

In several experiments, when proteins were incubated for long periods with pepsin and trypsin, the digests retained their nutritional value. However, it was still possible that some "special" molecule remained. And it had been repeatedly confirmed that neither mixtures of the amino acids that were then available nor the mixture of products from boiling proteins with mineral acids could support sustained growth or nitrogen retention in dogs or rodents.[29] Here, then, was a paradox: The evidence from digestibility studies tended to show that there was a complete breakdown to amino acids

25. Lea (1890), pp. 257–8.
26. Chittenden (1895), p. 113.
27. Abderhalden and Samuely (1905).
28. Conheim (1901), p. 462; Kutscher and Seemann (1902), p. 543; Matthews (1991), pp. 25–47.
29. Henriques and Hansen (1904), pp. 427–39; Abderhalden and Rona (1905), pp. 202–3.

before absorption, yet feeding studies indicated that amino acids could not replace protein.

Results obtained with sheep were exceptional, since adding a supplement of the amino acid asparagine, or even of ammonium acetate, to a low-protein diet improved their nitrogen balance. However, it was realized that the digestive apparatus of ruminating herbivores was different from that of single-stomach species such as humans, dogs and rats. The food eaten by these herbivores goes first to the rumen, a large bag in which the food is fermented for many hours before it passes to the true stomach, into which hydrochloric acid and pepsin are secreted. It was hypothesized that bacteria living and multiplying in the rumen were capable, like plants, of synthesizing protein from simple nitrogen sources and that dead bacteria passing down the gut might then be digested in the small intestine and utilized as with ordinary food protein. This idea was strengthened by Müller's finding that bacteria taken from a sheep's rumen could use asparagine and that the bacterial protein obtained was utilizable by dogs.[30]

The Discovery of Tryptophan

In 1872 Heinrich Ritthausen, who had spent many years identifying amino acids present in hydrolysates of plant proteins, concluded that the quantities of individual acids present differed from those in hydrolysates of animal albumins. He suggested that such differences might result in differences in the nutritional value of individual proteins.[31] Several more amino acids were discovered in protein hydrolysates during the next 30 years, and even the crude analytical methods available confirmed the differences in composition between animal and vegetable proteins.[32]

The discovery that some of the color reactions given by free amino acids in contact with particular chemical reagents were also given by intact proteins strengthened the idea that the amino acid "units" were already present in the protein rather than created by the procedures used to break the protein down. Because of a color reaction that they gave, most proteins were suspected to include an unknown amino acid, given the name "tryptophane," which contained an indole radical.[33] It was not found in acid hydrolysates but finally, in 1902, Hopkins and Cole at Cambridge University isolated it from a pancreatic digest of casein, after precipitation of its mercury salt.[34] (The name has been retained, but without the final "e.") It was found to be destroyed when heated with the strong acids used to hydrolyze proteins.

30. Müller (1906); Cathcart (1912), pp. 41–2.
31. Ritthausen (1872), p. 236; Chibnall (1939), p. 13.
32. Abderhalden (1905), pp. 21–3; Vickery and Schmidt (1931), p. 173.
33. Greenstein and Winitz (1961), Vol. 3, pp. 2317–21.
34. Hopkins and Cole (1902).

Four years later, Willcock and Hopkins carried out an important feeding experiment using mice. The basic diet had zein as its only protein source. Zein was an alcohol-soluble protein from maize that did not give the color reaction for tryptophan. Without supplementary tryptophan, the mice lost weight rapidly and survived on average only 16 days, but with the addition of this amino acid their weight remained almost constant and they survived for 30 days.[35] In one of a long series of experiments, Abderhalden found that the addition of tryptophan to an acid hydrolysate of casein was enough to make it capable of maintaining nitrogen balance in adult dogs.[36] These results certainly indicated that tryptophan had an essential role in animal biology but, as the authors pointed out, it was not necessarily involved in the synthesis of protein, since the animals did not gain weight. It could, for example, serve as a precursor to a hormone.

Eight years later, after Osborne and Mendel had developed protein-free vitamin sources that would support the growth of rats (i.e., protein-free milk and butterfat), they too did a feeding experiment with zein. They found that, when it was supplemented with both tryptophan and the amino acid lysine (which it also lacked), young rats doubled, or even tripled, their weight within 7 weeks. If only one amino acid was added they would not grow, but there was a difference between the responses. In the absence of tryptophan, body weight fell rapidly (as in the earlier mouse experiment), but in the absence of lysine, it remained almost constant for several weeks.[37] The authors suggested, first, that lysine was needed only for the synthesis of additional protein tissue. Second, it seemed possible that the "maintenance" requirement of adult animals for protein could not really be explained by "wear and tear" and a consequent need for synthesis of replacement protein. It could be that, in the main, tissue proteins were broken down when the intake was inadequate, simply to provide tryptophan for another function that had even higher priority.[38]

Another worker tested Liebig's old idea that vegetable proteins were used less efficiently because the animal consuming them had the additional work of synthesizing the creatine and other soluble nitrogenous compounds found only in animal tissues. He found that adding Liebig's "Extract of Meat" to the wheat protein gliadin did not improve its capacity to keep dogs in nitrogen equilibrium.[39]

The indications, therefore, all pointed to differences in amino acid composition as the explanation for young rats and mice not growing well when receiving only some vegetable proteins. However, there were also indications

35. Willcock and Hopkins (1906).
36. Abderhalden (1909).
37. Osborne and Mendel (1914), pp. 346–7.
38. Ibid., pp. 334, 340.
39. Michaud (1909), pp. 445–50, 456–7.

that some amino acids – specifically glycine – could be synthesized. In contrast, lysine and tryptophan apparently could not.

It seemed clear, as Rubner spelled out in 1911, that there could not be one value for the protein requirement of a human or animal; it must depend on the character of the protein.[40] But that "character" meant no more than "amino acid composition" was still no more than a hypothesis. Sorting this out was to require another 30 years of intensive work. On the chemical side, it required the development of more sensitive and reliable methods of amino analysis, as well as the search for additional compounds and their identification. It also required the development of standardized procedures for putting different protein sources on a scale of relative values, and we will begin there.

Measuring Protein Quality

The pioneers in the field were Osborne and Mendel, one of whose experiments has already been described. Like Atwater and Chittenden, they were both natives of Connecticut and educated at Yale, though only Mendel went on to spend a period of further training in Germany. Each then spent the remainder of his working life at a laboratory in New Haven and in each case married the daughter of his first chief (S. W. Johnson, director of the Connecticut Agricultural Experiment Station in the case of Osborne and Professor Chittenden of Yale's Department of Physiological Chemistry in the case of Mendel).

Thomas Osborne (born in 1859) was the older of the two. When the collaboration began, in 1909, he had already spent 30 years studying the characteristics of proteins isolated from different plant materials. He had made careful analyses of their contents of individual amino acids to the extent possible at the time and now wanted to see how far these values could be used to predict a protein's nutritional value. Lafayette Mendel (born in 1872), whose work had been mainly in the physiology of the digestion of foods, was pleased to collaborate in such studies, and in the next 20 years these two men were to publish well over 100 papers together.[41]

Their early experiments with growing rats, using diets consisting of purified materials, had shown them that a diet could appear to be adequate for a short period but then fail. They had, as described already, overcome this problem by including rich sources of vitamins, and they preferred to evaluate a diet by capacity to support rapid growth (measured as weight gain) and a healthy appearance over a period of several weeks, rather than by its capacity to support nitrogen retention over a few days in a balance

40. Rubner (1911), p. 46.
41. Smith (1956); Vickery (1956).

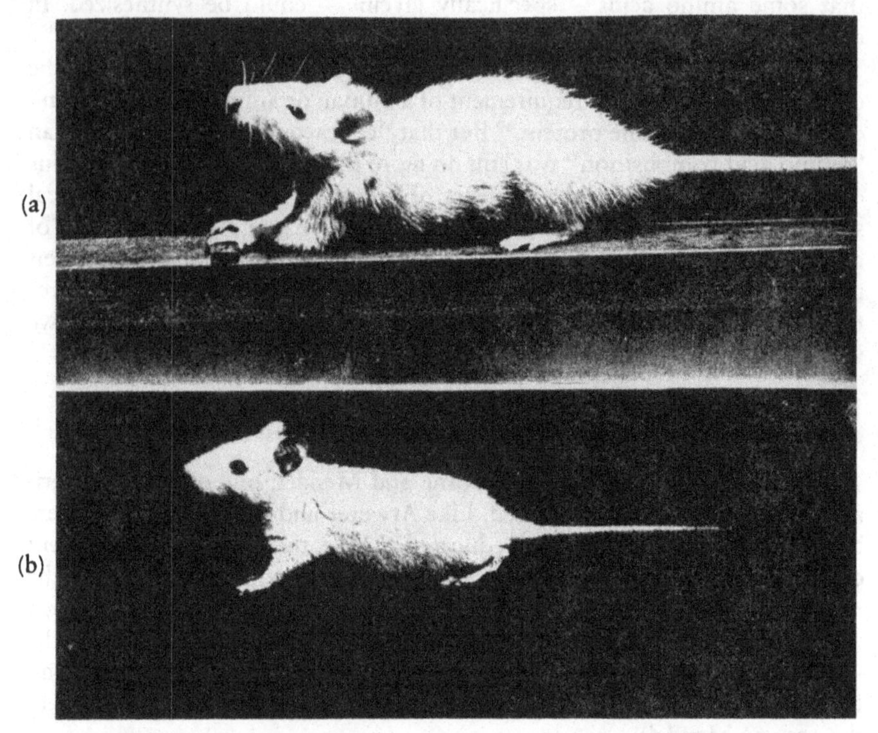

Figure 7.1. Comparison of two female rats, each 140 days old, having received (a) a diet containing 18% casein as the protein source and (b) the same diet but with 18% gliadin (from wheat) as the protein source. (Osborne and Mendel, 1911)

cage (Figure 7.1). One of their aims was to establish a standardized procedure that would allow workers to assign a reproducible numerical value to different individual proteins, or combinations of them. Controversies grew up around this subject. They do not need to be understood in order to follow the general thread of this book. However, the subject has had some practical importance, and the main developments, the definition of terms such as "biological value" and selected results are summarized in Appendix B.

Once satisfactory standardized methods had been developed, it became clear that proteins varied considerably in their capacity to support the growth of young rats. This remained true even after allowing for differences in the digestibility of the proteins. In general, plant proteins tended to be inferior to animal ones, though the earlier poor reputation of gelatin was confirmed.

Complementation between Proteins

In practice, human diets always include proteins from different foods. Workers were interested, therefore, in determining whether there were special advantages to be obtained from combinations of protein sources, particularly those used in traditional diets. The first experiments were carried out with mixtures of white flour and an animal food, with the white flour providing two-thirds of the total nitrogen, as well as with the two foods separately. Dried cow's milk fed as the sole source of protein gave a biological value of 84, and the flour fed similarly gave a value of 54. The arithmetic average for a 1:2 mixture of these materials is 64 (i.e., $\frac{1}{3} \times 84 + \frac{2}{3} \times 54$). However, the actual biological value of the mix was 14% greater than this. With whole eggs or meat, the "complementary" effect with white flour was of the same order.[42] Such an effect would be expected if the proteins in the mix had different essential amino acids in short supply. Obviously it was now hoped that it would be possible to relate the biological value of food proteins to their amino acid composition and thus to design new, and perhaps improved, complementary mixtures. But first it was necessary to know which of the amino acids were nutritionally essential, or "indispensable," an alternative adjective used in some papers.

Essential Amino Acids

The first important contributor to this part of the story is W. C. Rose, who did his graduate work under Lafayette Mendel at Yale and, after two relatively junior university appointments elsewhere, became professor of physiological chemistry (later changed to biochemistry) at the University of Illinois, Urbana, in 1922, at the age of 35.[43] Rose and Mitchell (whose work on protein quality is summarized in Appendix B), both regarded by their graduate students as admirable men, apparently had no personal contact during their working lives, although each had his laboratory on the Urbana campus. They finally appeared on the same platform when the University of Illinois organized a symposium in their joint honor, and they were both in their mid-seventies.[44]

Rose began a 20-year project to prepare a diet containing only free amino acids (rather than protein) that would support growth in rats and then to determine which were, or were not, dietary essentials. After a great deal of work in collaboration with chemists preparing sufficient quantities of 19 known amino acids, he began the feeding experiments in 1930. Although

42. Mitchell and Carman (1926); Block and Mitchell (1946), pp. 272–3.
43. Roe (1981).
44. Edman, Forbes and Johnson (1968); Roe (1981).

he included them all in his first trials, his rats failed to grow, as had those of his predecessors using fewer amino acids.[45]

With hindsight, it seems strange that he failed to include in his initial mixture the amino acid methionine, which had been discovered and identified as a component of proteins in 1923.[46] Its discovery had actually come from efforts to isolate an unknown growth factor for bacteria. But even though the work had been done in a department of bacteriology, the results had been published in the same publication – the *Journal of Biological Chemistry* – that was reporting nearly all the U.S. work on the nutritional requirements of rats. Finally in 1932, another laboratory indicated that methionine gave as good a growth response as cystine when it was added to a rat diet known to respond to the latter amino acid.[47] It was assumed that methionine, since it too contained sulfur, must be converted by rats to cystine, which had been classified as "essential" since the earliest work on amino acid supplementation. In later work, Rose fortunately included methionine, though regarding it even in 1936 only as an alternative to cystine.[48]

The next development was that he and his colleagues fractionated protein hydrolysates and found two that gave a growth response when fed together with their "amino acid diet." Further fractionation showed that the active factor in one fraction was isoleucine (a well-known amino acid and already in the basal mix but at a level well below that needed for growth). The other active factor was isolated and found, after further work, to be an amino acid not previously considered to have any significance for nutrition. It was to be named threonine.[49] Rose had achieved his first aim and could also work out exactly which molecules were or were not necessary for rat growth. Figure 7.2 is taken from Rose et al.'s final paper.[50] All but glycine can exist in either a "left-handed" (L) or "right-handed" (D) form. This is explained in any textbook of organic chemistry. Naturally occurring compounds (i.e., those synthesized in living tissues) are all in the left-handed form. When amino acids are produced by ordinary chemical synthesis, they are mixtures of equal parts of the two forms, the so-called racemic product. In some cases – for example, methionine – the right-handed D form can be converted in the animal body to the usable L form, and the cheaper DL-methionine mixture is often used to balance commercial poultry diets. However, with lysine, the D form is nutritionally useless, and it is the natural L form, produced by fermentation, which is sold commercially as a supplement.

45. Rose (1931).
46. Mueller (1923).
47. Jackson and Block (1932).
48. Rose (1936).
49. McCoy, Meyer and Rose (1935).
50. Rose, Oesterling and Womack (1948).

Classification of Amino Acids with Respect to Their Growth Effects in the Rat

Essential	Non-essential
Lysine	Glycine
Tryptophan	Alanine
Histidine	Serine
Phenylalanine	Cystine*
Leucine	Tyrosine†
Isoleucine	Aspartic acid
Threonine	Glutamic " ‡
Methionine	Proline‡
Valine	Hydroxyproline
Arginine§	Citrulline

* Cystine can replace about one-sixth of the methionine requirement, but has no growth effect in the *absence* of methionine.

† Tyrosine can replace about one-half of the phenylalanine requirement, but has no growth effect in the *absence* of phenylalanine.

‡ Glutamic acid and proline can serve individually as rather ineffective substitutes for arginine in the diet. This property is not shared by hydroxyproline.

§ Arginine can be synthesized by the rat, but not at a sufficiently rapid rate to meet the demands of *maximum* growth. Its classification, therefore, as essential or non-essential is purely a matter of definition.

Figure 7.2. Reproduction of the final summary of the rat's requirements for amino acids, as determined by Rose and his colleagues in 1948.

Amino Acid Analyses

It seemed clear that nutritionists now had to think not of "protein nutrition" but of "amino acid nutrition," since it was the amino acids alone which tissues received from digested food and which they needed, and that the whole array of 12 essential amino acids was needed for an organism to remain healthy.

In the nineteenth century there were great advances in chemical analysis, but virtually all the methods depended finally on weighing a product – usually an insoluble compound formed by reaction of the compound of interest with some other chemical. Analysis of amino acids was particularly difficult. Hydrolysis of proteins yielded a mix of up to 20 amino acids, many having very similar chemical and physical properties. In 1910 Osborne lamented the slow progress being made in this field and the difficulty of accounting for much more than one-half of the material present in a hydrolysate. Even a mix of known amino acids gave only 66% recovery with the separation methods available.[51] Thirty-five years later, the advances were

51. Osborne and Jones (1910).

reviewed in a monograph listing more than 700 references. Some amino acids were still being isolated and weighed, but for others specific color reactions were being used, with the depth of color obtained being taken as a measure of the amount of a particular amino acid present.[52]

In the 1940s a new approach was being developed, stemming from the work of microbiologists interested in developing purified media for growing lactic bacteria. These bacteria, living normally in milk that is uniformly rich in nutrients, have lost most of the powers of synthesis characteristic of other bacteria and have nearly all the nutrient requirements of the higher animals. Lactic bacteria were first used to assay materials for the content of the B vitamin riboflavin; the medium employed was designed to support rapid growth only when the vitamin was added. The rate of multiplication or acid production was then found to be proportional to the dose of vitamin added at suboptimal levels.[53] By 1943 it was realized that lactobacilli could similarly be used to assay an individual amino acid by using a medium containing all the other growth factors required, including a mix of all the amino acids except the one being tested.[54] The quantities required, as compared with those used in either traditional analytical methods or rat experiments, were tiny. Good responses could be measured in a small test tube after 48 hours of incubation. The expense was also small, and a large number of assays could be run at any one time.

By 1946 Block and Mitchell, in a now-classic review, related the amino acid composition of 23 protein-containing foods to their biological value determined with rats.[55] As a starting point, they hypothesized that the amino acid composition of whole hens' eggs, designed by nature for the synthesis of animal tissue, was ideal, and the level of each essential amino acid in another food's protein was expressed as a percentage of the corresponding value for egg protein. It was judged that the amino acid present at the lowest level (in proportion to egg) would be the first-limiting amino acid for the synthesis of growing animal tissue. Thus if a particular protein had only 60% of the lysine content of egg protein and this was the lowest percentage value, then 10 g of this protein would result in the synthesis of 6 g of mixed animal protein, but then there would be no lysine left and synthesis would stop. The excesses of the other amino acids would then be metabolized to provide urea plus carbon skeletons or carbon dioxide and energy.

Originally such a protein was described as having a "40% deficit," but it later became the usual practice to refer instead to it having a "chemical score" of 60% or just 60. A plot of the original 23 sets of results showed

52. Block and Bolling (1945).
53. Snell (1979).
54. Mitchell and Block (1946).
55. Block and Mitchell (1946), pp. 260–7.

a clear tendency for the materials with a higher chemical score to have a higher biological value. However, there was not an exact correspondence. To use the statistician's term, there was a "correlation of 0.86" (compared with 1.0 if all the values were to fit exactly onto a straight line when plotted in the same way). This means that we can expect, if we know that the chemical score of a sample is 50, that the corresponding biological value will be within 7 units of 70 – that is, to be in the range from 63 to 77 – and that 1 sample in 20 might be twice as far from the best estimate of 70. The plot also showed that even a sample with a zero chemical score might still be expected to have a positive biological value.

The last point is explained by the apparent capacity of rats to maintain nitrogen balance with a protein completely lacking lysine, this being the limiting amino acid for growth for many of the samples. Other factors adding variability to the plot are inherent inaccuracies in the analyses, as well as variability in rat responses. Finally, the amino acid pattern of egg proteins intended for the growth of chicks may not contain exactly the minimal requirement of each amino acid for optimal utilization by rats.

Despite these shortcomings, it was possible to confirm for a good number of foods that the amino acid calculated to be first-limiting was in fact the only supplement that gave the food an improved biological value when it was tested with rats. It was also calculated for several mixtures of two foods that showed a complementary effect, as discussed earlier in the chapter, that this could have been predicted from their amino acid analyses. To return to the example of 1 part of milk protein and 2 parts of white flour protein, the first-limiting amino acid in white flour was calculated to be lysine, giving it a score of only 28, and that in milk to be methionine plus cystine, with a score of 68. The score for the mixture was 53, though by simple proportion it would have been only 41.[56] In popular writing, the protein in a vegetable food is sometimes described as "incomplete." This term is misleading, since all the amino acids are present, but not in ideal proportions.

Studies with Humans

Obviously, workers were eager to know how far the results obtained with rats were relevant to human nutrition. To determine the complete list of amino acids essential for humans by direct experiment required much larger quantities of L-amino acids than were needed for the work with rats, but Rose was able to publish the results in 1949 in preliminary form.[57] Eight of the 10 amino acids found to be essential for maximal growth in young

56. Ibid., pp. 272–3.
57. Rose (1949).

rats were also found to be required by adult men. The exceptions were histidine and arginine. Further work on the quantitative requirements of adults for amino acids was carried out after 1950 and will be considered in Chapter 11.

It had been relatively easy to carry out nitrogen balance trials of the type pioneered by Thomas using protein foods as the sources of nitrogen. Unfortunately, the results in the 1920s and 1930s continued to be contradictory. This was thought to be due in part to adaptation periods being too short and to total energy intakes not being controlled.[58] Mitchell, using his long experience in nitrogen balance work, collaborated on a carefully controlled study published in 1945.[59] Nine women students served as subjects, energy intakes were controlled to maintain body weight and each protein food, or mix, was fed at five different levels. It was estimated that the proteins of cow's milk had a biological value of 74, and the proteins of white flour a value of 41. Both of these are lower than the values of 84 and 54 obtained with rats for the same materials; however, the relative proportions are similar. The mixture of proteins in a typical U.S. diet had an intermediate biological value of 65 and 90% digestibility. The mean daily requirement of the mix for nitrogen balance was estimated to be 50 g crude protein for someone weighing 70 kg, after making an allowance for what the authors described as "adult growth," including losses of nitrogen from skin, hair and sweat. Similar results were obtained in the next study, with a diet in which one-third of the protein came from meat and the remainder from mixed plant sources. When the meat was replaced by an equivalent level of protein from white bread, 23% more total protein was required, on average, to achieve balance.

The results from the more carefully designed human studies seemed to indicate that measures of protein quality that had been obtained with rats would prove applicable to humans as well.[60] Yet the quantities of protein needed to produce nitrogen balance in human adults, even with relatively poor quality protein foods, were lower than the protein intakes in the practical dietaries that investigators had studied. There was therefore no obvious practical problem of protein deficiency, and a common saying among nutritionists at the end of the period was: "Look after the calories and the protein will look after itself."

This did not apply to subjects who had suffered severe traumas such as bone fractures, major surgery, certain types of infections or extensive burns. They went into a state of shock, which resulted in greatly increased losses

58. Leitch and Duckworth (1937).
59. Bricker, Mitchell and Kinsman (1945), pp. 278–9.
60. Ibid., pp. 273–8; Hegsted, Tsongas, Abbott and Stare (1946), pp. 272–9.

of urinary nitrogen.[61] It was believed that paying special attention to their subsequent intake of protein aided greatly in their recovery.

There was concern, during the Great Depression in Western countries in the 1930s, that low-income or unemployed workers and their families were malnourished because they could not afford a balanced diet. However, the conclusion from the best-known study from this period was that the common deficiencies were of vitamin A and calcium rather than of protein.[62] After World War II, when food was still scarce in Germany, studies in two orphanages showed that 9- to 10-year-old children thrived and made up for deficiencies in both height and weight when they received as much bread as they wished to eat, together with vitamin and mineral supplements, but less than 9 g of animal protein per day. In fact, they did just as well as other children receiving three times as much animal protein.[63]

The Dynamic State of Body Protein

Meanwhile, the use of new techniques based on developments in physics and chemistry during the 1920s and 1930s were beginning to have a revolutionary effect on biologists' ideas about the behavior of the constituents of the body. These new techniques involved the use of isotopes. It had come to be realized that ordinary nitrogen (to take an example) is actually a mixture of atoms having an atomic weight of 14 and a small proportion (0.38%) having a weight of 15. The ^{15}N atoms, as they are called, are *isotopes* of nitrogen; they have an additional neutron in their nucleus but the same pattern of electrons in orbit around the nucleus. It is the latter that gives the isotope the same chemical and biological properties as ordinary ^{14}N.[64]

Similarly, ordinary hydrogen contains a small proportion, less than 1 atom in 10,000, of "heavy hydrogen," or "deuterium," with an additional neutron in its nucleus and an atomic weight of 2 instead of 1. Once physicists and chemists had discovered ways of accurately measuring the ratios of the different isotopes present in a particular sample, and then of preparing materials with high proportions of the same isotope, biological experiments could begin. Fortunately it appeared that living organisms treated these isotopes exactly as they did the more abundant forms of hydrogen and nitrogen.

The scientist whose name is particularly associated with the early work

61. Cuthbertson (1931); Cuthbertson (1942); Lund and Levenson (1948).
62. Orr (1937), p. 41.
63. Widdowson and McCance (1954), pp. 52–9.
64. Clarke (1948), pp. 3–15.

in this field is Rudolf Schoenheimer. He came to Columbia University in New York in 1933 as a Jewish refugee from Nazi Germany. Columbia was the leading center for the separation of isotopes at that time, and he was able to collaborate there with David Rittenberg and others experienced with these isotopes. Tragically, Schoenheimer was to commit suicide in 1941, when still only 43 years old, but by then their work had totally changed people's perception of the body's metabolism.[65]

We will discuss one experiment in enough detail to give an idea of the approach, though not of the work and technical expertise required.[66] A quantity of the amino acid leucine was prepared with a high proportion of the nitrogen atoms labeled with ^{15}N and the hydrogen atoms in the side chain labeled with deuterium. This was then added to the diet of four mature rats for a period of 3 days at a level equivalent to that present in the protein of the diet (15% casein). The rats were then killed and their tissues analyzed. From the deuterium content of the leucine isolated from the rat carcass protein, it was concluded that 6.7% of it had come from the leucine consumed in the previous 72 hours; this amounted to 32% of the leucine consumed.

This was in striking contrast to the ideas of protein metabolism held by nutritionists. They had found that the endogenous loss of urinary nitrogen (i.e., the loss on a nitrogen-free diet with adequate energy intake) was approximately 58 mg/day for rats of the size (325 g body weight) used by Schoenheimer.[67] The body protein of such a rat contains approximately 8,300 mg nitrogen. The daily urinary loss therefore corresponds to 0.7% of the body's store of protein nitrogen. It had therefore been thought that this corresponded to the daily "wear and tear" of body protein and that for an animal to remain in nitrogen balance the same quantity of new protein would have to be synthesized each day. In the 3 days of Schoenheimer's trial this would correspond to 2.1% of body protein. And some workers had believed that the breakdown (and need for replacement) would actually be less if the animals were in a better nutritional state.

However, Schoenheimer's results indicated that more than three times this amount of leucine was incorporated into body tissues. He postulated that the peptide chains of body proteins opened and reclosed, allowing individual amino acids to escape and be replaced by others. It was a difficult concept to accept. One suggestion was that the newly arrived leucine might be lodged in a kind of temporary storage material rather than in true "working" body components. Schoenheimer and his colleagues countered this with

65. Ratner (1979); Guggenheim (1991).
66. Schoenheimer, Ratner and Rittenberg (1939).
67. Smuts (1935), pp. 419–21.

a demonstration of the rapid labeling of a specific antibody protein in the blood at a time when its concentration in the blood was actually declining.[68]

The same research group also demonstrated that the rapid appearance of labeling in body proteins was not peculiar to leucine. Moreover, when the ^{15}N label was followed, it seemed clear that different amino acids were actively exchanging amino groups. Thus when leucine was fed, only one-third of the ^{15}N found in the rats' body proteins was still attached to leucine; the remainder was attached to other amino acids, most notably glutamic acid. Only the lysine remained entirely free of ^{15}N. All these findings were to be confirmed and extended by other workers.

These were disturbing results. Work on the metabolism of body fat had similarly startled biochemists as to the *Dynamic State of Body Constituents*, the title of Schoenheimer's final monograph. As he wrote, "If the starting materials are available, all chemical reactions which the animal is capable of are carried out continually," and at another point, "The synthesis of amino acids, like that of fatty acids . . . proceeds even where there is no obvious need for it."[69]

Schoenheimer did not envisage that whole proteins were being dismantled but, rather, that proteins while still in position in the tissues opened up at particular points to release an individual amino acid, which was rapidly replaced by another molecule of the same compound. To use an analogy, it was as if one visited Venice and saw builders swarming over the face of every building, chipping out stones, throwing them into boats drifting by on the canals and replacing them with exactly comparable stones from the next boat that came along. Then to complete the analogy one would have to imagine there being one particular part of the city where any building stones that drifted by were broken into rubble and sent to an external dump, so that the city required a continual supply of new building stones arriving from outside, in exactly the same condition as those being systematically destroyed elsewhere in the system.

The picture indeed boggles the mind, and we will return to it in Chapter 11. One recalls Claude Bernard's criticism of balance studies made some 80 years earlier, when he said that one could not hope to understand what went on inside a house just by analyzing what went in the doors and what came out of the chimney.[70] And the new developments were certainly a blow to the self-respect of nutritionists, who had felt that because they had discovered nutrient requirements and interrelationships (such as methionine being able to replace cystine, but not vice versa) they were at the cutting edge of biochemistry.

68. Schoenheimer (1942), p. 36.
69. Ibid., pp. 31, 39.
70. Holmes (1964), p. xcviii.

Mitchell was one of the few nutritionists even to discuss the significance of the findings of Schoenheimer and his successors.[71] He accepted the importance of the work, but argued that there must still be some unknown regulating system at work that allowed organisms to come into nitrogen balance on widely different intakes. Although it could no longer be assumed that all of the extra nitrogen that appeared in the urine of someone on a high-protein diet came directly from digested amino acids, the reality of determined nutritional requirements was unchanged and a net nitrogen balance was still a measure of dietary adequacy of protein for adults.

Summing Up

In the 40 years covered by this chapter many things were explained, but a new mystery developed. The earlier failure of dogs to thrive indefinitely on some low-protein diets was explained by the diets being deficient in the newly discovered trace nutrients – vitamins. The failure of early amino acid mixtures to replace proteins in nutrition was also explained. In the case of acid hydrolysates of good-quality protein, it was due to tryptophan having been destroyed, and in work with mixtures of purified amino acids it was due to their lacking methionine and threonine. It now seemed clear, at least from animal experiments, that the true nutritional requirements were for amino acids rather than protein and that, when vitamins were also supplied, the level of protein did not need to be high provided that it contained a reasonably good balance of the essential amino acids. Many individual isolated proteins (such as gelatin and zein) were of low biological value, and those of whole vegetable foods also tended to be somewhat poorer nutritionally than those of whole animal foods, but the combinations present in ordinary mixed diets were at least of adequate quality in all the examples studied. The newer studies also indicated that nitrogen balance was obtained at levels of protein intake below even those recommended by Chittenden, and there seemed no reason to be concerned about protein being deficient in human diets.

Mendel put it eloquently in 1923: "The recognition of the relative importance of the nitrogenous foods for nutrition . . . brought about a glorification of the albuminous substance . . . which has persisted in its extreme form almost until the present time." Now, he added, "the pendulum of enthusiasm about the proteins has swung from one extreme to the other."[72]

The new mystery was the unexpectedly dynamic state of body constituents, as found in the work with isotopes. Specifically with regard to proteins,

71. Mitchell (1942), pp. 257–62.
72. Mendel (1923), p. 14.

amino acids were apparently constantly leaving and reentering tissue proteins. The fact that amino acids leaving protein were then reusable made it difficult to continue explaining the adult need for amino acids by the old concept of "wear and tear."

8 Protein Deficiency as a Third World Problem, 1933–1957

THE *ARCHIVES OF THE DISEASES OF CHILDHOOD* for 1933 included a short article that described a serious disease, and apparently a form of malnutrition, in the Gold Coast (then a British colony, but now Ghana) in West Africa. The author said that it appeared to be distinct from any of the usual deficiency diseases.[1] The only response to this article was a note in the next volume of the journal from H. S. Stannus, the recognized British authority on deficiency diseases in the tropics, saying that the author was in error and that, from its description, the disease was clearly the infantile form of pellagra.[2] Nevertheless, 50 years later the *Archives* reprinted the article in full, under the editorial comment that it was perhaps the most important article the journal had ever published, and much of the work to be covered in this and the following chapters can be seen to have stemmed from it.

KWASHIORKOR

The author of the article was Cicely Williams, an Englishwoman born in 1893 who had grown up in Jamaica and received her medical training in Oxford and London. Jobs were scarce at that time for young women doctors, and she applied to the Colonial Medical Service, hoping to be posted to Jamaica.[3] After a 2-year delay she was appointed in 1929 to a position in West Africa – still a dangerous place at that time for Westerners without resistance to yellow fever and other endemic diseases. Her special interest was in infants and young children, and she was disturbed by the high death rate among them. One disease, in particular, the subject of the article already referred to, seemed to occur almost entirely in children 1 to 4 years old

1. Williams (1933).
2. Stannus (1934), pp. 733–4.
3. Dally (1968); Darby (1973); Craddock (1983).

who had been weaned early (by local standards) and then fed entirely on gruel made from white corn (maize) after some degree of fermentation. Those who were not seriously ill generally recovered if given cod-liver oil and canned condensed milk as a supplement. Although goats were kept locally, they were not milked. It was not customary for anyone to drink animal milk.

In her introduction Williams said that the disease was characterized by "edema [swelling from accumulated fluid], chiefly of the hands and feet, followed by wasting, diarrhea, irritability, sores (chiefly of the mucous membranes), and desquamation [peeling off] of areas of the skin in a manner and distribution which is constant and unique.... It appears to be due to some dietetic deficiency and to be uniformly fatal unless treated early."

She then described, as a typical case, that of an 18-month-old child. First there was a slight edema; then patches of skin became "dark, thickened and crumpled" in a sort of "crazy pavement" pattern. As the older patches matured they stripped off, leaving a pink, raw surface. The child was now in a state of misery, had persistent diarrhea and would die in a few more days if untreated. Later photographs of children suffering from the condition and from marasmus ("simple" starvation) are reproduced here (Figure 8.1). Hurried postmortem examinations carried out under difficult conditions showed nothing abnormal except fatty livers.[4] Finally, Williams apologized for a lack of references to published literature, explaining that she had no access to a library.

In 1935 she published a paper in the *Lancet* based on her experience with 60 further cases.[5] Here she used for the first time the local name for the disease – *kwashiorkor* – which a nurse from the local community told her meant "the sickness that the older child gets when the next child is born."[6] Stannus again issued a quick rejoinder, criticizing Williams for failing to accept his earlier conclusion that the disease was infantile pellagra and also for choosing to introduce a native name for the condition.[7]

Was It Pellagra?

What was the significance of this dispute? First, if the condition were really a known disease, then the proven forms of treatment should be adequate. Second, since intensive research was in progress in the United States on the identification of the pellagra-preventive factor using the dog as a model animal, there would be no reason for a separate study of the factor needed

4. Williams (1973), 337; Craddock (1983), pp. 63–4.
5. Williams (1935).
6. Dally (1968), p. 59.
7. Stannus (1935).

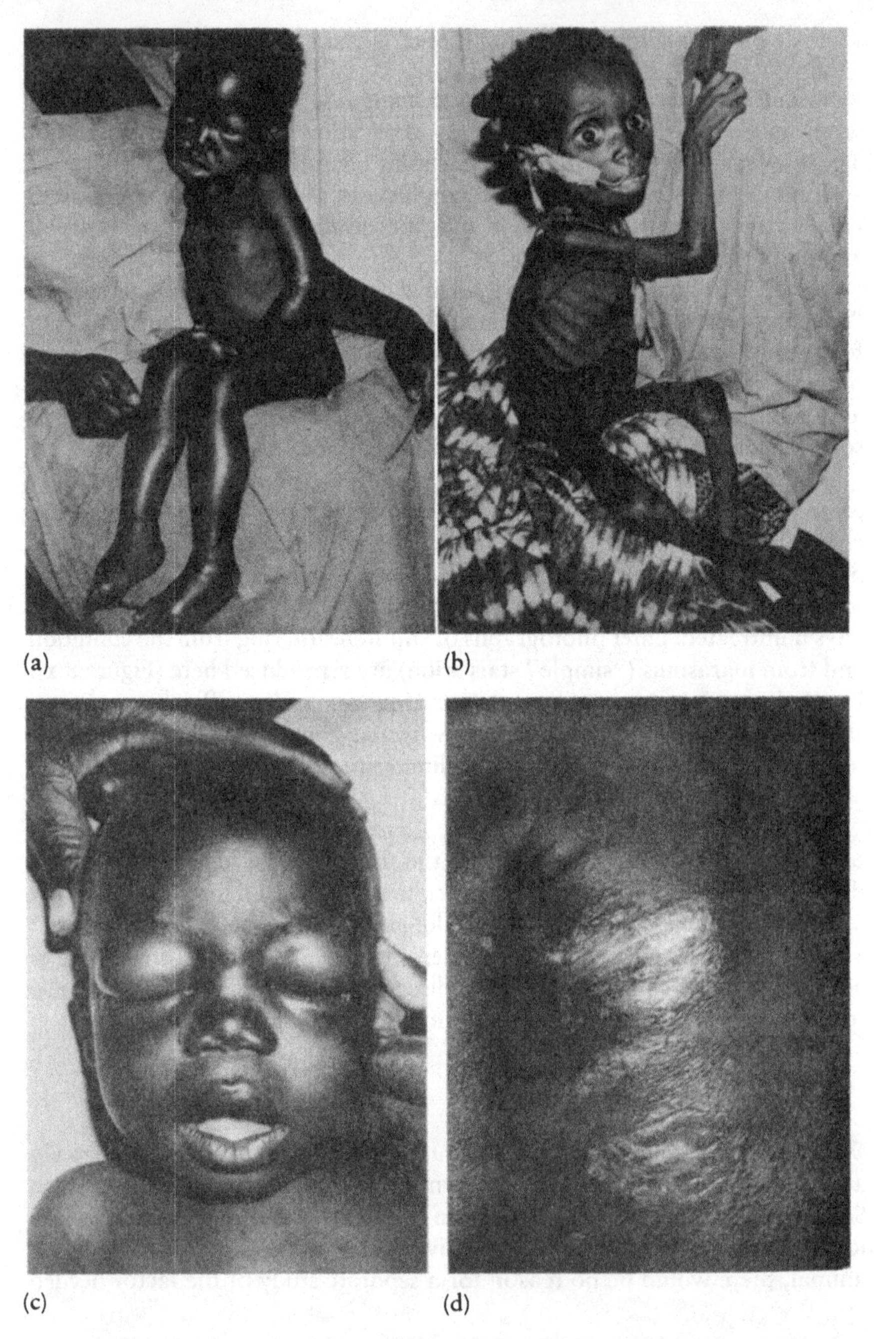

Figure 8.1. (a) An African child with kwashiorkor; (b) another with marasmus ("simple" starvation) for contrast; (c) edema of the face of a child with kwashiorkor; (d) skin lesions typical of kwashiorkor. (Dr. R. G. Whitehead)

to prevent the development of kwashiorkor. Cicely Williams's connection of the disease with a corn diet must have strengthened the idea that it was a form of pellagra, since pellagra's association with corn had been seen in so many other parts of the world.[8]

A similar controversy had arisen in the case of another deficiency disease, scurvy. Barlow's disease in infants showed some differences from adult scurvy, and many physicians had therefore refused to consider it essentially the same disease despite its occurrence with the same type of foods as were associated with adult scurvy. But in the 1920s, with better histological studies of the changed tissues, the condition was accepted as infantile scurvy.[9] Although Stannus's dogmatic and even contemptuous reaction may be partly explained by the alleged identification of kwashiorkor as a new disease having come from a doctor inexperienced in tropical work, and from a woman, the identification of the disease with infantile pellagra was made by others in Africa as well as himself.

More constructively, another medical officer on the Gold Coast published a note in the *Transactions of the Royal Society of Tropical Medicine and Hygiene* in 1935, saying that he had seen a case exactly like those described by Dr. Williams and that, although its origin was still obscure, he entirely agreed that it was "impossible to confuse it with any of the well-known and often described diseases."[10] This journal circulated widely in the British tropical colonies and a response came, within the same year, from a medical officer in Kenya, saying that an identical condition had attracted attention there in the past two years, though pallor of the skin and hair changes seemed to be more constant features than had been seen by Williams. The majority view was that "some form of food deficiency is the root of the trouble."[11] There was no reference to the possibility of the disease being a form of pellagra.

We can now see from literature reviews that cases with the characteristics of kwashiorkor had been reported in at least 11 other papers by this time from different parts of the world and in different languages, but without their being linked.[12] At this point Cicely Williams found herself unexpectedly transferred to Malaya, where she was to remain through World War II and to endure great hardship in Japanese prison camps.[13] But the disease that she had been so interested in was, from then on, to receive intensive study, particularly in East and South Africa.

Hugo Trowell, an English physician working in Kenya in the early 1930s,

8. Harris (1919), pp. 33–132.
9. Carpenter (1986a), pp. 166–72.
10. Sharp (1935).
11. Carman (1935).
12. Trowell, Davies and Dean (1954), pp. 2–8.
13. Craddock (1983), pp. 105–23.

wrote, much later, that medical missionaries were already familiar with the disease and had provisionally thought of it as "protein deficiency" because of the edema. However, they and Trowell himself had accepted Stannus's apparently authoritative statement that it was infantile pellagra. In 1935, when Trowell was transferred to Uganda, he saw many more cases, which were described there as "congenital syphilis."[14]

Trowell published a review of the work done up to 1940 under the title *Infantile Pellagra.* He cited papers indicating that the same disease occurred in Mexico and Costa Rica as was seen in Africa (including the Belgian Congo, as it was then). One had called it pellagra-beriberi, attributing the skin changes to pellagra and the edema to beriberi (recognized as being caused by thiamine deficiency). However, he had injected thiamine and not obtained any reduction of edema.[15] He therefore suggested that protein deficiency caused the edema (as it did under wartime conditions in adults), but that the other changes were a form of pellagra, which had been shown in the United States to respond to the vitamin niacin (nicotinic acid).

Trowell injected seven sick children with niacin. Of these, two died and two more were removed from the hospital by their parents after they had shown no rapid improvement, but the three remaining all showed significant improvement after 5 days. All the children also received in this period "as much milk as they could take, an egg each day, and a little bread in addition to the usual hospital diet of steamed plantains [starchy bananas]." Trowell's tentative conclusion was that "the major deficiency is one of acute pellagra associated probably with nutritional edema.... treatment should be by nicotinic acid injections, blood transfusions, marmite [yeast extract] and a diet rich in protein."[16]

Williams replied from Malaya to Trowell's article in a rather conciliatory way, agreeing that the East African cases were essentially the same as those she had seen and going on to say: "What's in a name? It matters very little as long as we understand each other." Nevertheless, the African children did not display the same type of skin changes as those that she had, by then, had the opportunity to see in New Orleans, where there had been a few cases of pellagra in children; their skin changes were all on the areas exposed to sun, as in adults. Williams was willing to accept that individuals could be suffering from complex deficiencies but concluded that, until much more was known of the biochemistry of these things, "it is unwise and confusing to the issue to label a variety of symptoms with a name that implies an as yet unproved identity."[17]

14. Trowell, Davies and Dean (1954), p. 18.
15. Trowell (1940), pp. 389–98.
16. Ibid., pp. 398–403.
17. Williams (1940).

Trowell and Muwazi in 1945 used the new term "malignant malnutrition" for the disease.[18] They explained that they did not like the term "kwashiorkor" because they understood that it actually meant "red boy" in the Ga language of the Gold Coast, and although a change in hair color had been noticed it was probably specific to Africans and not a necessary condition for diagnosis of the disease.[19] However, it was agreed later that the word really did mean "the deposed child."[20]

The authors had seen the condition in 48 children from 7 months to 3 years of age. Almost all the children in the region of Kampala, the Ugandan capital, had contracted malaria by their third month, and hookworm disease was also common after the second year. "Malignant malnutrition" was seen mostly in the local Ganda tribes, whose babies were not weaned until well into their second year, and then restricted to three meals per day of cooked plantains, sweet potatoes and tea.[21] The absence of corn in the diet of these children may also have influenced the authors' decision to give up describing the disease as a form of pellagra.

Another striking finding was the greater number of cases seen among adults than among children. Most of the 144 adult sufferers were from the Belgian mandated territory of Ruanda-Urundu and had walked 500 to 800 miles to obtain work in the plantations around Kampala. They had been short of food on their journey and, after arrival, lived typically on sweet potatoes together with very small amounts of meat and vegetables. It was estimated that they had daily intakes of approximately 2,000 kcal, 25 g protein, 1 g fat, 250 mg calcium and 8 mg iron. They had very low body weight, edema, "crazy pavement" dermatosis and diarrhea, but not always pallor of the skin and hair. All autopsies showed fatty livers. Both the children and adults had abnormally low levels of albumin in the blood serum, but increased globulin levels. The adults responded, but only very slowly, to a standard hospital diet of plantains, meat, beans and potatoes, supplemented by liver, milk, thiamine and nicotinic acid.[22] That the adult patients did not have the appearance of pellagrins may have been another reason for Trowell's change of name for the infantile disease.

Failures with Vitamins

In the same year, 1945, another paper appeared, this time from Johannesburg, reporting more cases, still described as "infantile pellagra" but characterized by edema. The authors carried out liver biopsies and reported that

18. Trowell and Muwazi (1945), p. 241.
19. Ibid., p. 242.
20. Trowell, Davies and Dean (1954), p. 11.
21. Trowell and Muwazi (1945), p. 231.
22. Ibid., pp. 230–9.

fatty infiltration of the liver was a constant and early feature of the condition. They treated seven patients with combinations of nicotinic acid, riboflavin and thiamine, but all died. Another six patients were treated with 10 g/day of "Ventriculin," a proprietary, dried extract of hog's stomach, and within a week they all showed a spectacular recovery.[23] Further trials with the same batch of Ventriculin were also highly successful, and Trowell, too, reported successes.[24] Unfortunately, later trials with stomach extracts were much less successful, and even a later batch of Ventriculin appeared relatively ineffective.[25] No explanation has been found for these apparently contradictory results.[26]

In 1946 a medical officer in Southern Rhodesia (now Zimbabwe) reported further cases. He considered that the cause could not be "dietetic" since the disease did not occur during the season of greatest food scarcity and was seen in children whose diet differed in no way from those who remained healthy. His suggestion was that the central problem was liver damage from some toxic factor in certain batches of corn and cassava.[27]

From this period on, papers on the subject began to appear in increasing numbers.[28] In the next two years, it was reported from South America that the disease could occur in white children as well as people of African stock. There was general agreement as to the value of milk and to the opinion that this was due to its protein fraction. It was also suggested that damage to the pancreas might precede liver damage and that treatment of the disease should be directed to restoring pancreatic function, since this was judged to be the organ most damaged.[29]

In 1949 Trowell again reviewed the subject. He still believed that it was a complex syndrome: "The strands of the weft are . . . total calories, protein, vitamin B complex; the warp . . . many different strands of an infective nature, helminths, pneumonia and so forth." Even apparently trivial infections had to be treated rigorously. "All sources of animal protein should be high. Milk should be at least one pint." Dietary protein should always be suspected as the bottleneck to recovery if edema was severe and serum albumin levels very low, and the latter was the most constant feature of the condition. "To correct this we give powdered milk, like a medicinal powder several times a day. . . . all vitamins are forbidden, so are unnatural foods like dehydrated liver, desiccated stomach, proteolysed meats, casein hydrolysates. Everything must be directed to overcoming the false economy which would de-

23. Gillman and Gillman (1945).
24. Trowell (1946).
25. Gillman and Gillman (1951), pp. 301–2.
26. Trowell, Davies and Dean (1954), p. 182.
27. Gelfland (1946).
28. Trowell, Davies and Dean (1954), pp. 1–46.
29. Davies (1948); Waterlow (1948a); Waterlow (1948b).

prive wards of essential diets but would flood them with expensive vitamins and quack foods."[30]

INVESTIGATION BY THE UNITED NATIONS

After World War II, the new United Nations Organization (UNO) set up subsidiary agencies, including the World Health Organization (WHO), with its headquarters in Geneva, and the Food and Agriculture Organization (FAO), in Rome. In October 1949 representatives of these two agencies met to coordinate their plans for nutritional work. First on the agenda was endemic goiter and the practical problems of supplying iodine preparations in ways that would prevent its appearance in deficient areas. Second on the list was kwashiorkor. It was believed that cirrhosis (fibrous degeneration of the liver) found in many adults in Africa and Central America might be the result of their having suffered from the disease as children. It was important for WHO to extend its studies to areas where the disease did not occur but in which the diet was apparently similar to those where it did. There were no suggestions as to what nutritional deficiencies might be responsible for the condition, but it was urged that more animal milk should be made available for weanlings, or "where that was difficult, the production and use of foods and/or preparations which can act as a partial substitute for milk should be vigorously encouraged."[31]

Dr. J. F. Brock, professor of medicine at Cape Town, and Dr. M. Autret (on the staff of FAO) toured Africa in 1950 to investigate the occurrence of kwashiorkor and the different methods used for its treatment.[32] They reported that essentially the same disease was a widespread and serious one among infants in many areas, though they did not make any estimate of its actual prevalence. Growth retardation, edema and low serum protein levels were the most constant factors. Dyspigmentation, dermatosis and diarrhea were all variable. Livers were uniformly found to be fatty, but not always enlarged. Mortality was very high unless the condition was treated. It seemed inappropriate to call the condition "malignant malnutrition," since it had been treated successfully with skimmed milk.

Brock and Autret concluded that both the epidemiology (i.e., freedom from the disease in areas where meat, fish and/or milk were relatively abundant) and the successes with skimmed milk and some good-quality vegetable protein sources pointed to protein deficiency as the cause of the disease. However, it was still possible that the missing factors were particular amino acids or a vitamin such as vitamin B_{12} that occurred in association with

30. Trowell (1949).
31. Joint FAO/WHO Expert Committee on Nutrition (1950), pp. 15–16.
32. Brock and Autret (1952), pp. 32–4.

these proteins. Methionine was a possible limiting factor since a deficiency of this amino acid led to fatty deposition in the livers of experimental animals.

Despite the proven value of skim milk, it was impractical to expect milk production to increase rapidly in Africa, and imported material would be too expensive for general use, though it was felt that skim-milk powder should be made available to medical centers for use in intensive treatment. Substitutes for milk were required, and these would have to be mostly vegetable products: cereals in place of cassava, which was particularly low in protein, and legumes (peas and beans) acceptable to African people. Studies were required of the ability of infants to digest different protein sources and of the effects of traditional processing that involved fermentation or germination of cereal foods.

Effects of Protein Deficiency in Adults

Brock and Autret also made a separate point. Two adult diseases were unusually common in the areas studied – cirrhosis and primary carcinoma (cancer) of the liver. It was thought that parasitic infestations could be responsible in some areas. However, in the Johannesburg area these diseases were common even though parasitism was not a major problem. Since kwashiorkor, characterized by liver damage, now seemed to be caused by a dietary deficiency of protein, it seemed likely that adult cirrhosis was also the result of prolonged protein deficiency, though the two diseases were not necessarily directly related to each other. Kwashiorkor therefore pointed to a much wider problem: "It would not be too far-fetched to attribute to protein deficiency, at least in part, the backwardness of the African people.... Protein deficiency, though alleviated after the post-weaning phase of life, appears at least in a part of the African population to be permanent throughout life... causing a great volume of premature morbidity and mortality."

According to Brock and Autret, it was difficult to estimate the nutrient content of African diets because quantitative data were not available. "Accordingly, a unit of 100 calories has been adopted for comparing different foods." (Actually, the authors should have written "kcal" instead of "calories" because they were referring to the energy required to raise the temperature of 1 kg water by 1°C.) They compared the concentration of protein in different foods with the estimated requirement for protein (in the same units) at different ages. The maximum protein requirement was given as 4.5 g protein per 100 kcal for children 2 to 3 years old.[33] Since the time of this report, it has become more common to convert such values to a simple

33. Ibid., pp. 56–8.

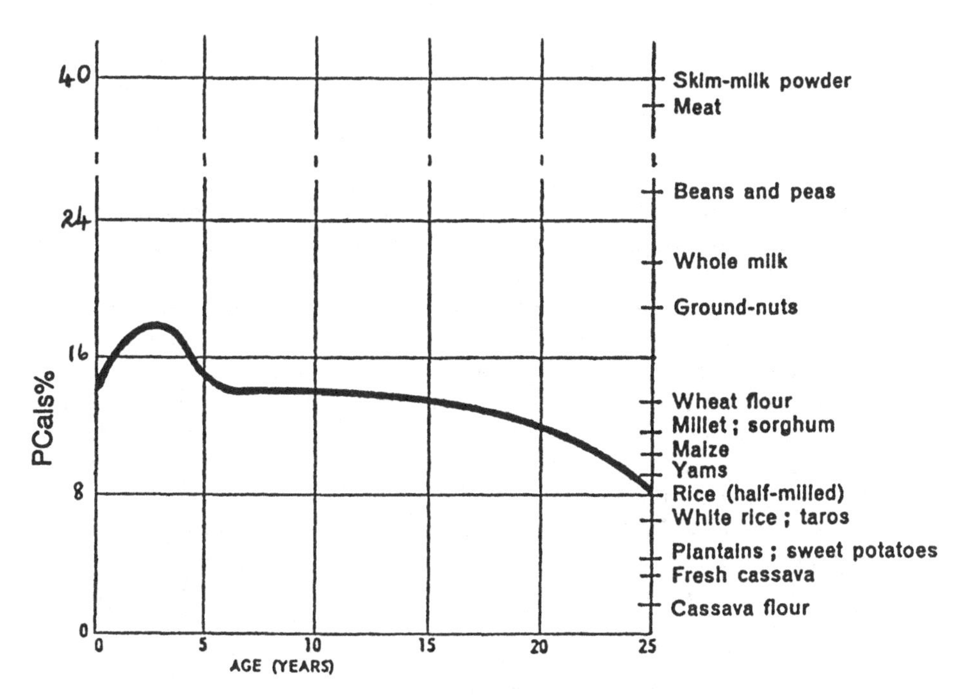

Figure 8.2. Brock and Autret's 1952 diagram relating the protein concentration in foods to human needs at different ages for average-quality protein, all per 100 kcal with the protein units changed from grams to kilocalories to give PCals%.

percentage by assuming that dietary protein has a usable energy equivalent of 4 kcal/g. Then 4.5 g is equivalent to 18 kcal, and the requirement for 2- to 3-year-olds becomes "18 PCals%," meaning that 18% of the usable energy in the diet must come from protein. The corresponding value for 20-year-olds at 3 g protein per 100 kcal was "12 PCals%." The authors' diagram is reproduced here in these units (Figure 8.2). It was clear that plantains, sweet potatoes and cassava, with only 0.5 to 1 g protein per 100 kcal (2–4 PCals%), were grossly deficient in protein. Even the cereal grains (wheat, rice, corn) would be judged inadequate for the youngsters. Where plantains were the staple food, the diet as a whole could reach the standards for protein only if it included considerable quantities of foods rich in protein, such as fish, meat, milk and beans.

The authors also estimated the supplementary food requirements of a 12-month-old African child receiving only 86% of its calorie requirements from breast milk with 2.2 g protein per 100 kcal (i.e., 8.8 PCals%). Their calculations were as follows:

	kcal	Protein (g)
Daily requirement of the child	730	20
Supplied by breast milk	630	14
Deficit	100	6

For 100 kcal of supplementary food to supply the full protein deficit it would have to have at least 6 g protein per 100 kcal (or 24 Pcals%), that is, the level in beans or equal parts of skim-milk powder and cereal.

Using the same values, one would find that, even if the 12-month-old child were to receive enough breast milk to satisfy his or her full energy requirements, this would provide only 16.2 g protein, which is short of the requirement by 3.8 g. It seems strange that human milk should itself appear "unbalanced" even when consumed at a level that meets calculated energy demands in full, but the authors were forced to choose standards for protein of "average quality," and milk protein is clearly "above average."

The joint FAO/WHO Expert Committee met for its second session in 1951. Brock and Autret's draft report was accepted. It was agreed that milk production in Africa could not be increased significantly in the short term and that the use of imported dry skim milk could be only a temporary measure. Policymakers must stress the provision of locally produced foods. In some areas it should be possible for fish production to be increased rapidly. More information was needed on the kinds of pulse and vegetable combinations that were effective in preventing kwashiorkor, and obtaining such information should receive priority in the allocation of research funds. The portion of the report that dealt with liver disease in adults and its likely association with protein deficiency was not, it appears, discussed in any detail. However, the committee noted, "Prevention of the syndrome calls for a general improvement in the diet of the whole population."[34]

Autret and Moise Behar (a research worker in Guatemala and consultant to WHO) then made a regional study in Central America. There, too, a disease equivalent to kwashiorkor in its essentials was widespread, regardless of race. Within the area it was called *sindrome policarencial infantil* (the multiple deficiency condition of infants). Helminthics (drugs used to treat worm infestations) and purges were commonly thought to precipitate the disease, because the infant's resistance to the side effects of the drugs was so low.[35] One difference between the situation in Central America and that in Africa was that in the former region 40% of the cases were seen in children over 4 years old. In general, giving vitamins had had a bad effect. Milk, and especially skim milk, had proved the most effective for treatment.

34. Joint FAO/WHO Expert Committee on Nutrition (1951), pp. 19, 22–8.
35. Autret and Behar (1954).

Dietary studies had indicated that intakes of calories as well as of protein were inadequate and that both should be corrected. The following records for three 3-year-olds in the initial stage of the syndrome illustrate the point (RDAs indicate recommended daily allowances of the U.S. National Research Council):[36]

	Actual intakes (% of allowance)		
RDAs	Child 1	Child 2	Child 3
kcal (1,200)	44	61	37
Protein, g (40)	45	33	38

The Value of Vegetable Proteins

In 1951 the British research group at Kampala was joined by R. F. A. Dean, who had been a member of the Medical Research Council's team that studied diets in German orphanages after World War II. He had obtained very promising results in the orphanages, even among infants, with mixtures of cooked soybeans and malted cereals as the main sources of both energy and protein.[37] In Uganda he began with milk powders to provide a standard of comparison, but found that severely ill children continued to have diarrhea for long periods when given a large quantity (170 g) of dry skim milk daily. He thought that this was caused by the large amount of lactose being consumed. Further tests used a more concentrated source of milk protein, and better results were obtained when "Casilan," containing 80% milk protein, replaced part of the milk powder. When this mixture was given to severely sick patients, the death rate dropped from approximately 20% to 10%. Diarrhea cleared up more quickly if the milk powders were mixed with either cooked plantains or mashed sweet bananas, and these mixtures were also more palatable. The usual aim of all these treatments was to provide 50 to 75 g protein and 1,000 kcal per day, which had a balance of 20 to 30 PCals%.[38]

Dean also carried out trials with soybeans. The dry beans were soaked in water for 2 days, rubbed to remove the husks, then minced, packed into jars with water and boiled in water for 10 hours, a procedure designed to remove their bitter flavor. The resulting "mush" was mixed with the other ingredients of the test diet in a blender to the consistency of a thick cream. The first mixes had included a proportion of corn flour. However, such a

36. Ibid., p. 45.
37. Dean (1949); Dean (1953a), pp. 63–110, 115–28.
38. Dean (1952); Dean (1953b), pp. 769–70.

large volume of water was needed to bring the mix to a suitable consistency that the children could not take in enough nutrients. Sick children had a greatly reduced appetite.

A mix that proved more successful consisted of 300 g cooked soy, 750 g sweet banana, 75 g sucrose, 8 g calcium lactate and 3.5 g of a vitamin mix. The vitamins included 15 micrograms (μg) cobalamin (vitamin B_{12}). With this mixture, children recovered from kwashiorkor, though not quite as rapidly as with the mix based on skim-milk powder and Casilan. The oil in the soy appeared to be better digested than milk fat. Soy protein had been found in rat feeding experiments to require supplementation with methionine in order to support optimal growth, but Dean felt that slower-growing infants would not have such a high requirement. Methionine, together with cystine, the second sulfur-containing amino acid, was needed in relatively large amounts by young rats because of their growth of fur, which has a high cystine content.

The practical problem was that soybeans, though grown in large quantities in some countries bordering the East African coast, were not a common crop in Uganda. Sunflowers were grown, and although the seeds required more processing to produce an edible flour, it seemed possible to simplify the procedure. The final problem of acceptance of a new food might have been exaggerated, because there was no problem with the acceptance of a can of condensed milk by people who had the money to pay for it.[39] This work was described and discussed in more detail in the book *Kwashiorkor* that appeared in 1954.[40] It gives a good summary of knowledge up to 1952.

While this work was still in progress, the FAO/WHO Expert Committee held its third session at the end of 1952, this time in West Africa, and "concentrated its attention on protein deficiency and its effect on child health."[41] The condition was reported in an increasing number of countries. In India, there was a condition called "nutritional dystrophy" or "malnutritional oedema syndrome" in many 1- to 3-year-olds living on rice gruel and little or no milk.[42] In some the condition was complicated by signs of vitamin A deficiency, which might have been due, at least in part, to poor absorption of fat. The children typically had diarrhea and heavy roundworm infestation, and also showed extreme apathy.

The committee's report listed local names for diseases resembling kwashiorkor – for example, "infantile edema" (Guatemala), "farinaceous dystrophy" (Uruguay), "red baby" (French Cameroons) and "sugar baby" (Jamaica) – and put them all under the heading "Names for Protein Mal-

39. Ibid., pp. 771–81.
40. Trowell, Davies and Dean (1954), pp. 195–202.
41. Joint FAO/WHO Expert Committee on Nutrition (1953), p. 4.
42. Achar (1950); Gopalan and Patwardhan (1951).

nutrition in Children." The introduction to the report stated that it was concerned with "disease in children associated with deficiency of protein in the diet" and "the prevention of protein malnutrition through appropriate developments in food production... a problem of fundamental importance throughout the world." There was also reference, as before, to the likelihood that low-protein diets would reduce the ability of men and women to engage in sustained work.

On the assumption that it was protein deficiency specifically that was causing the problem, more definitive recommendations could be made as to desirable changes in food patterns. Cassava (also called manioc or tapioca), which had originally been brought to Africa as a famine (i.e., reserve) crop, was becoming accepted as a staple, replacing traditional grain crops. This was undesirable since it (like plantains) had only 2 to 3% of its calories in the form of protein, while grains had 8 to 12%. The report stated that breast feeding should be prolonged where high-protein weaning foods such as cow's milk were not available, and that parents should think of their weanlings as needing "special" foods rather than having to rely on what they could manage to consume from the adult diet.

The committee warned against technical experts coming in from other cultures and making recommendations before they had learned existing customs and beliefs; they should be helpful in their attitude rather than authoritarian and superior. More research was required on the quantitative study of diets associated with kwashiorkor, on the nutritive value of the foods available in problem areas and on the means of detecting malnutrition in its early stages. Kwashiorkor was distinguished from *marasmus,* which was regarded as the result of simple starvation, or calorie deficiency. This led to a gross loss of both muscular tissue and subcutaneous fat so that the victims had a shrunken look, in complete contrast to the swollen, plump-looking kwashiorkor victims. Hair changes and edema were rarely seen and the sufferers maintained their appetite. If marasmus was the result primarily of calorie deficiency, then kwashiorkor must presumably have a different cause, that is, a deficiency of protein. It was recognized, however, that there were intermediate conditions with some signs of one disease and some of the other – for these the term "marasmic kwashiorkor" was sometimes used.

In 1953 Waterlow (from Jamaica) and Vergara (from Chile) carried out a study in Brazil, though the report was not published until 1956.[43] The authors saw numerous cases of the disease known in Brazil as *distrofia pluricarencial* and judged it to be essentially kwashiorkor. Skin and hair changes were not as common as those reported in Africa. The condition was described in the title of the report and elsewhere in the text as "protein

43. Waterlow and Vergara (1956).

malnutrition," though it was stated that "the intake of calories is also frequently inadequate." In certain regions of the country one-third of the children appeared undernourished. Most of these children were from city slums and in their second year. It was considered of interest that the disease occurred in white children as well as in those of other races. Serum protein levels were very low, but fatty liver was a variable feature and therefore was not the essential lesion. Cirrhosis and primary carcinoma of the liver were rare in Brazil. There was no confirmation, therefore, of the idea, based on the African experience, that these diseases in adults might be long-term sequelae of kwashiorkor and of continuing protein malnutrition.

In 1953 a conference was held in Jamaica, with 26 participants, from every continent.[44] The title of the conference was *Protein Malnutrition,* though several members doubted whether the disease(s) under discussion could be described in this way. Scrimshaw said that "in INCAP [the Institute of Nutrition for Central America and Panama] we are far from convinced that the biochemical changes encountered are specific to protein malnutrition; rather, they seem to be non-specific." Another participant said, "The hypothesis of protein malnutrition in kwashiorkor, fertile though it looks, has not yet been proved in terms of either qualitative or quantitative depletion of protein." However, there was general agreement that "what we see clinically are cases in which the dominant theme is protein malnutrition, and that there are local variations in signs and symptoms brought about by an overplay or a variation in the theme, caused by deficiencies of one or more of half a dozen nutrients."[45]

Gopalan, from India, said that in his experience the diet of children in poor communities was low in calories. Dean, too, said that children in Uganda who showed signs of kwashiorkor had quite unexpectedly low caloric intakes. Even the "sugar babies" in Jamaica, who had been thought to have a high calorie intake, were reported by Katerina Rhodes to lose their appetite some 2 to 4 weeks before developing kwashiorkor.[46]

Waterlow raised the paradoxical point that marasmus, in which both calories and protein were equally deficient, seemed easier to cure than kwashiorkor, in which there was thought to be a better supply of calories (usually in the form of carbohydrate). Yet it was known that normally a low calorie intake led to more rapid depletion of protein. There was reason to think that the kwashiorkor state needed to be triggered by some stimulus, such as an infection.[47]

The idea that adult cirrhosis of the liver followed from kwashiorkor was

44. Waterlow (1955).
45. Ibid., pp. 4, 82, 158.
46. Ibid., pp. 43, 179, 250.
47. Ibid., pp. 8, 16, 48.

agreed to be no longer tenable since, in addition to its rarity in Brazil, a high incidence of cirrhosis had been found in parts of Africa where there was no kwashiorkor and meat was relatively abundant.[48]

The participants in the conference agreed that the digestive enzymes of the sick children were grossly deficient, mainly as a result of the abnormal state of the pancreas, and this made them unable to tolerate fat. Skim milk had proved about the best treatment everywhere. However, there was agreement, as at an earlier conference, that locally available plant protein sources would have to be developed. Work in both India and Mexico had confirmed that giving victims vegetable mixes based on beans could result in their cure (and therefore could presumably be used also for prevention), though recovery was usually slower than with mixes containing milk protein. One report to the conference described the acceptance by schoolchildren of bread made with a proportion of fish flour, but others commented that 6-month-old infants might react with diarrhea to a food tolerated by older children.[49]

In late 1954 the Committee on Nutrition met again in Geneva to review the main nutrition problems. "Protein malnutrition" was first on the agenda, with a review of measures being taken, or under consideration, for supplementary feeding programs. During 1953 enough dried skim milk to provide about 9 g protein per head daily had been distributed through UNICEF (United Nations International Children's Emergency Fund) to 6,000 children in the Belgian colonies in Central Africa, and it was judged that the condition of the children had improved. It was hoped that after these supplies ceased "the protein needs of the people will be met by the extension of fish farming."[50]

In 1955 another meeting was held in the United States, with the title *Human Protein Requirements and Their Fulfilment in Practice.*[51] Participants were puzzled by reports of kwashiorkor appearing occasionally in breast-fed infants. Recent analyses had indicated that milk secreted by women in conditions of poverty still had a normal concentration of protein. Scrimshaw (from INCAP) also referred to cases developing suddenly in children although there had been no change in their diet. A stress factor such as infection was thought to be responsible.[52] Senecal (French West Africa) and Hansen (Cape Town) raised a new point – that kwashiorkor patients were as depleted of potassium as they were of protein and that giving potassium salts was important in their recovery.[53]

48. Ibid., p. 109.
49. Ibid., pp. 207, 220, 253–7.
50. Joint FAO/WHO Expert Committee on Nutrition (1955), pp. 7–8; Holemans, Lambrecht and Martin (1956), p. 5.
51. Waterlow and Stephen (1957).
52. Ibid., pp. 85–7.
53. Ibid., pp. 91–2.

With regard to protein requirements, it was agreed that "the allowances recommended by previous committees, and based mainly on intakes in Western countries, are at all ages too high, and to reduce them is not only realistic but scientifically sound." Allowances of 100 g protein per day for adults were thought unrealistic, when even nursing mothers were lactating successfully with 70 g or less, and there was no evidence of adolescents being at risk from protein deficiency during their "growth spurt" period.[54] Information on the requirements of both infants and adults for individual amino acids was reviewed. It was agreed that where the protein in a diet had a poor amino acid balance, more of it would be needed, but it was not yet possible to calculate Third World diets in terms of each separate amino acid; the data were inadequate.[55]

With regard to protein supplies, Autret said that FAO's first principle was to develop milk production wherever practicable: "The next approach is how to use for feeding children various protein-rich foods now used for animal feeding only. Of one million tons of fishmeal now being produced in the world each year, 30 to 40 per cent... could be used for feeding children. In Africa... 500,000 tons of peanut meal are available, which might well be turned into an edible protein food; there are also large quantities of soy flour." Gopalan commented: "The general premise seems to be that protein malnutrition in under-developed countries is directly and solely the result of the primary dietary inadequacy of protein.... A greater contribution to the prevention of protein malnutrition in certain regions of the world may lie, not in supplying supplements, but just in eradication of malaria, the improvement of poor home economic conditions, etc.... factors ... responsible for the precipitation of protein malnutrition."

There was debate at the meeting as to how much testing of different products would be needed before they could be officially recommended. While some urged the danger of using untried materials, and the possibility of long-term adverse effects, others said that children were continuing to die for lack of high-protein supplements and, where similar products had been fed for years to young chicks and pigs, it seemed unreasonable to hold them back from infants.[56]

In 1956 Professor Brock's group in Cape Town reported that they had given 33 infants with kwashiorkor a diet containing sugar, minerals, vitamins and mixtures of amino acids (but no protein as such). In 27 of the children, cure was definitely initiated. In some cases vitamins were withheld at first, and there was still a positive response.[57] The authors concluded it

54. Ibid., pp. 114–16, 133.
55. Ibid., pp. 105–7, 151–2.
56. Ibid., pp. 138–9, 160–7.
57. Hansen, Howe and Brock (1956).

was no longer necessary to suppose that an associated factor (such as an unknown vitamin) explained the value of milk and other high-protein foods in the treatment of kwashiorkor.

In 1957 the Committee on Nutrition met for the fifth time, though its composition had changed considerably over the years. With regard to protein malnutrition, the members reported that the first stage of their work had involved the collection of information about the extent of the condition and its importance as a public health problem. The second had been the technical analysis of the problem by experts. The third stage, now in progress, was the development of preventive measures, including the effective use of protein-rich foods other than milk. In late 1955 WHO had set up the Protein Advisory Group "to advise on the safety and suitability of proposed new protein-rich food preparations," and the Rockefeller Foundation had given $250,000 toward the work. UNICEF had also given $100,000 toward the procurement of materials and general support of the program.[58] The progress of work of this kind will be the subject of the next chapter.

Summing Up

In 1933 a fatal disease of young West African children was described for the first time in the English-language literature. Called "kwashiorkor," it was characterized by edematous swelling of the legs, peeling of the skin, an enlarged liver, great misery and loss of appetite. The sufferers had usually subsisted mainly on corn gruel since weaning. It was later realized that the same condition had already been described in other poor countries under a variety of names in French- and Spanish-language publications and was widely distributed.

In Africa the condition was first considered to be the result of deficiencies of B vitamins, but this was not confirmed. And since the condition was typically associated with the low-protein staple foods plantains and cassava, and could be treated successfully with reconstituted milk powders, it came to be classified as "protein malnutrition," though a further stress such as infection might be needed to precipitate the acute condition. It was realized that those affected had low intakes of calories as well as of protein. However, the clinical picture was quite different from that of marasmus, the condition of babies suffering from starvation, which was considered primarily a calorie deficiency. Children with kwashiorkor were also slower to respond to treatment.

The specialist organizations set up by the United Nations after World War II considered that correcting the problem should be given the highest

58. Joint FAO/WHO Expert Committee on Nutrition (1958).

priority. It was feared that low-protein diets in early childhood and continued into adulthood in the Third World resulted in impaired development that was never made up. There was particular need for alternative protein sources where it was not practical to expect a local dairy industry to increase milk supplies greatly.

Brock was to say, with some justification, in 1960 that "in human nutritional studies and in international public health this has been a protein decade,"[59] and another member of the FAO/WHO committee said, "We have moved from the era of vitamin research to protein research."[60] Autret wrote in 1962 that it was the assumption of the Nutrition Division in FAO that "deficiency of protein in the diet is the most serious and widespread problem in the world."[61]

59. Brock (1961), p. 1.
60. Sebrell et al. (1961), p. 541.
61. Autret (1962).

9 International Efforts to Produce High-Protein Supplements, 1955–1975

As we have seen, international committees of nutritional experts concluded in the period 1950–5 that substitutes for milk were urgently needed to solve the world's most critical nutrition problem – a shortage of protein. They set out to stimulate and organize the production of new kinds of high-protein materials that would help to fill the "protein gap."

At their fifth session, in 1957, the FAO/WHO experts set criteria for these materials. They had to be capable of being produced locally, to be affordable by all sections of the population and to have a long storage life even under tropical conditions. They also had to be acceptable, nontoxic and easily included in ordinary diets. Finally, their protein must effectively supplement existing diets.

Six protein sources were studied: fish, soybean, peanut, sesame, cottonseed and coconut. With regard to the last four, attention was focused on the low-fat press cakes produced by the vegetable-oil industry.

The initial step would be to produce a batch of each type of high-protein product under carefully defined conditions and then to study its biological value and safety with animals before testing its value for children and adults. The next step would be to study the possibilities of manufacturing the product in the area concerned. The program had already reached this point with two of the products in question – fish flour and *saridele* (a soy extract).

The experts concluded that, although such foods could not permanently solve the problem of meeting the protein requirements of vulnerable groups, because in underdeveloped countries transportation, processing and distribution tended to make these products relatively costly, they could be expected to make an important contribution. And although the ultimate aim was to teach mothers to feed their children adequately by giving them sufficient foods of the right kind, this could be effective only if the necessary foods were available.[1]

1. Joint FAO/WHO Expert Committee on Nutrition (1958), pp. 22–3.

The U.S. National Research Council had recently established the Committee on Protein Malnutrition, which could submit applications for research money and, in turn, distribute it to particular projects, in collaboration with WHO's Protein Advisory Group (PAG).[2] In 1961 the PAG formally became a joint responsibility of FAO and UNICEF as well as of WHO. The original members of the group were clinical nutritionists. To this group were added food scientists, technologists associated with the development and manufacture of protein foods and marketing experts.

Despite this, the FAO/WHO Expert Committee on Nutrition, meeting in 1966, said that "the marketing phase of the [protein-rich food] program, including consumer studies and promotion campaigns, has not been given the attention it requires.... Industrial production of protein concentrates ...has been slow and uncertain.... Plans should be developed in co-operation with existing distribution organizations."[3]

In the late 1960s the problem of the "world protein gap" came to the attention of the UN headquarters staff in New York. The Advisory Committee on the Application of Science and Technology to Development recruited a panel of experts to advise them as to what should be done about the problem. Its report was accepted and published in 1968 as *International Action to Avert the Impending Protein Crisis.*[4] The report stated: "The protein gap in the nutrition of the population of our planet is becoming a most important scientific, technological and public health problem and a national and international policy issue." The panel recognized that difficulties had arisen with many of the schemes designed to attack the problem, some of them stemming from neglect of human motivation and behavior, others from a lack of basic training among the personnel assigned to them.[5]

The panel's list of specific proposals included the development of fish protein concentrate and oilseeds, as in the 1957 proposals listed earlier, and they added the following: the development of genetically improved plants; greatly intensified research on "single-cell protein sources" (i.e., yeasts and bacteria); and the use of synthetic amino acids to fortify and improve the nutritive value of dietary proteins. They also recommended the development of research and training centers and schemes.[6]

The estimated cost of implementing these proposals was $75 million, but they were well received by the General Assembly and the UN Secretary General, and another booklet elaborating the points in more detail was published in 1971.[7] It was agreed that the work of the PAG should be

2. Darby (1975), pp. xv–xvii.
3. Joint FAO/WHO Expert Committee on Nutrition (1967), pp. 60–1.
4. Advisory Committee on the Application of Science and Technology to Development (1968).
5. Ibid., pp. 40, 59.
6. Ibid., pp. 29–31.
7. Panel of Experts on the Protein Problem Confronting Developing Countries (1971).

broadened; the World Bank became an additional sponsor, and sociologists and economists were added to the group. In 1974 the name was changed to the Protein–Calorie Advisory Group to the United Nations System.[8] This reflected a changing view of the nature of kwashiorkor, which will be discussed in the following chapter, and in 1977 the work of the Group was terminated.[9]

Many projects to produce high-protein materials that were started in different parts of the world within the period 1956–77 were independent of the PAG, but its meetings and quarterly bulletins did provide a forum for discussing the problems and progress in this field.

Fish Protein Concentrate

The first project that FAO and UNICEF were working on at the time the PAG was established was a "fish flour" plant in Chile. Later such material came to be called "fish protein concentrate" (FPC). The attraction of this project is easy to understand. The Chilean government was concerned about the nutritional state of young children from the poorest segment of the country. It was proving difficult and expensive to expand the dairy industry significantly, and importing dry skim milk was an economic drain. Yet Chile had a long coastline, and considerable quantities of fish were already being caught for conversion to a dry powder (i.e., fish meal) that had proved, in many parts of the world, to be a valuable source of supplementary protein for chickens and pigs and could be produced more economically than milk powders. Surely, with some small modifications, an upgraded flour could be prepared for young children, who should have a higher priority than animals for such a source of good-quality protein.[10]

The great bulk of the world's fish meal in the 1950s was made from clupeiform species such as herrings and anchovies. They are generally small, less than 12 inches long when mature. They also live in "schools," comprising as many as 1 million or more fish, which keep tightly together. Though often moving hundreds of miles over a season, they characteristically stay close to shore. When a school moves into their area, fishermen can catch enormous quantities relatively cheaply. Some are sold whole, while others have their heads and guts removed and are either smoked or canned. But when very large catches are landed, not all of the fish can be processed before they deteriorate. Fish meal plants were therefore developed and used to process large portions of the catch – particularly in Scandinavia, in South Africa and on the coasts of Chile and Peru.

8. Anonymous (1974).
9. Schatan (1977).
10. Pariser, Wallerstein, Corkery and Brown (1978), pp. 144–7.

Clupeiform fish have a relatively high content of fat, which is distributed throughout their tissues. If the whole fish were dried, the product would contain 30% fat or more. This would be unacceptable for animal feeding. The high fat content would give a "fishy" flavor to the carcasses or to hens' eggs. Also, the fat (or oil) would quickly oxidize when in contact with air, making the meal indigestible and possibly toxic. At the same time the heat evolved during the oxidation could be sufficient to set a whole store of fish meal on fire.

The processor of clupeiform fish therefore removes most of the oil by first steaming the fish to soften and break up the tissues, and then pressing out the liquor from them. The hot liquor is then centrifuged, and the oil can be sold separately. The aqueous layer can be returned to the "press cake" (i.e., the solids), and the mix dried in a current of hot air and ground to a powder. The resulting meal may have some 7% fat, which is acceptable for animal feeding if relatively low levels of the meal are included in the animals' diets.

The fish flour first tested in Chile to determine whether it could be introduced in foods such as bread without consumers noticing or objecting had come from South Africa. There, ordinary fish meal made from clupeiform fish had been extracted with hot ethanol (ethyl alcohol) and the residual solvent removed by heating in a vacuum oven. This reduced the fat content to below 1%. It still had a slightly fishy flavor, but this was undetectable when added as 4% of the dry ingredients of brown bread. Tests with rats showed no loss in the value of the protein and no evidence of toxicity, though the process removed a large proportion of the water-soluble B vitamins from the meal. More than 800 tons of fish flour were incorporated into brown bread in South Africa at government expense in the period from 1956 to 1959, but the project was stopped because it was realized that those at risk from protein deficiency (i.e., preschool children of the poorest nonwhites) did not consume a significant amount of bread.[11]

With the success of the small tasting trials of the South African product in Chile, in which enriched bread was given to 140 children each school day for 6 weeks, the UN agencies agreed to assist in setting up a fish flour plant there.[12] The goal was to produce an essentially fat-free flour with no fishy flavor. The plant was set up to process hake, a lean white fish in low demand. The fish were first dried in the usual way and then extracted, first with hexane and then ethanol. The first trial runs in 1958 were unsuccessful, but by mid-1961 1 ton of FPC had been produced. When further difficulties were encountered with the hexane extraction, the Chilean government decided, in view of the continuing problems in developing an economical

11. National Nutrition Research Institute (1959), pp. 59–61, 144–7.
12. Roels (1972), pp. 5–6.

source of FPC, to drop the program and concentrate instead on increasing the supply of milk powders for children.[13]

At this time there were plans to produce FPC in a number of countries.[14] We will follow that of the Bureau of Commercial Fisheries in the United States, whose history has already been analyzed in an important book.[15] In 1961 John F. Kennedy, just elected as president of the United States, was anxious to demonstrate the country's technological abilities – mainly in the space race, where the USSR had so far been leading – but also in other ways. The Bureau put forward the idea of developing FPC as a new form of "food from the sea," which would help the U.S. fishing industry and also provide technical expertise for the Third World to adopt in turn (Figure 9.1).

Research was begun in the Bureau's laboratory, adjoining the campus of the University of Maryland. In 1963 an experimental model plant was built that would process meal with no fishy flavor from a fish that again had little fat. It was thought prudent to begin with hake, which would be easier to process than the more abundant fatty fish. The procedure chosen consisted of extracting the wet, chopped-up fish with hot isopropanol (isopropyl alcohol), a chemical generally similar to ethyl alcohol but a better solvent of fat. This was designed to remove fat and water at the same time, with the solvent being recovered fairly cheaply by distillation. The principles of this procedure had already been worked out in a Canadian government laboratory.[16]

Under political pressure to produce rapid results, and before the problems of economic, continuous extraction had been worked out, the Bureau agreed to ask for funds to build a commercial plant as a pilot for private industry. In April 1971 the plant, set up on the West Coast, was ready for operation and 1,500 tons of hake were processed in the next few months. However, the hake resource, which was thought to have been abundant, failed. If the plant was to work with a reasonable output, it would therefore have to be modified, at considerable expense, to operate with fatty fish. But the prices of fuel and of all species of fish were rising rapidly, and Congress no longer seemed willing to believe that FPC would be economically attractive. No private company was interested in taking over the plant, and it was sold for scrap. The detailed histories of FPC projects in other countries were different, but the final outcome was the same.

The Bureau also had problems meeting the strict specifications of the Food and Drug Administration (FDA) for a human food product. The first

13. Pariser, Wallerstein, Corkery and Brown (1978), pp. 148–51.
14. Roels (1972), pp. 3–11; Pariser, Wallerstein, Corkery and Brown (1978), pp. 117–43, 161–213.
15. Pariser, Wallerstein, Corkery and Brown (1978), pp. 19–113.
16. Idler (1969).

More people need more protein

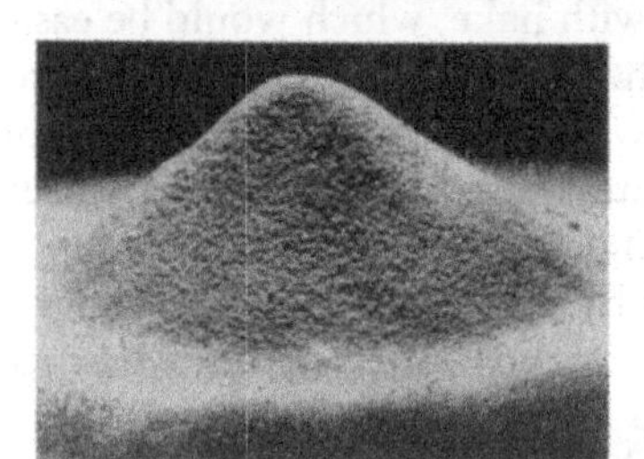

FPC

FISH PROTEIN CONCENTRATE

IT'S FOR PEOPLE

The amount of fish now caught is sufficient to provide nearly an ounce of fish daily to everyone in the world; however, because of uneven distribution, many people living inland get little or none. It is believed that the fish catch could be increased threefold and thus make an important contribution to the protein shortage, but the problem is to preserve the fish in an economical manner so that it can be shipped, stored, and put into other foods cheaply and safely.

Fish protein concentrate, known as FPC, can provide an answer. It is made by removing the water and oil from ground fish, leaving just the protein and the minerals in a concentrated form. FPC is a light, neutral powder containing about 80 percent high quality protein. It may be added to many foods without changing their appearance or flavor, yet greatly increasing and improving their protein value.

Figure 9.1. Material from a 1971 U.S. Department of Commerce booklet entitled "FPC – Fish Protein Concentrate: It's for People" (1971).

such criterion was that FPC be free of disease-causing bacteria. This meant that the processing plant had to be built to satisfy the hygiene rules laid down in the United States for any human food product. Since farm animals appear to be less sensitive to bacterial food poisoning, the same precautions are not required for the production of animal feedstuffs and therefore of fish meal. The second was the requirement that foods be free of "filth." The public would expect the FDA to ensure that, say, ground beef for a hamburger was not contaminated with cow dung or with intestinal contents. Yet when thousands of small clupeiform fish were being processed, as in the South African procedure, the whole fish was used, including the guts and gut contents. The cooking killed the bacteria present – and fresh sea fish, in any case, did not carry bacteria that cause disease in humans – but there remained the aesthetic concept of filth, and the dairy industry objected that FPC represented unfair competition for milk powders, which had to meet the most stringent hygienic standards. Finally, it was ruled that FPC could be sold only in 1-lb. packages, so that purchasers would be buying a clearly labeled product. This eliminated the use of FPC as one ingredient in proprietary mixed foods or formulas, where consumers might not realize what they were getting.[17]

Another problem arose from the bone content of dried fish. The major mineral in bone was calcium phosphate, and this was generally regarded as a nutritional asset. However, fish bones also contained a certain amount of fluoride. This was again desirable in the human diet, *up to a point,* because it strengthened tooth enamel and increased resistance to decay. But at higher levels it could cause mottling of the teeth and, at extremely high levels, rigidity of the spine. The FDA required that the level of fluoride in FPC be kept below 80 parts per million. This could be achieved only by removing a proportion of the bones, either before processing or by sieving the partly ground product. Either approach added to the cost of production.[18]

The Bureau had assumed that it would be politically impossible to export to Third World countries, even as free gifts to the needy, material that had been considered by another U.S. government agency to be unacceptable human food. It was also believed that the product would have to be virtually tasteless in order to be acceptable and that this would mean almost complete extraction of its fat. Some people argued that this was unnecessary, and a fishy flavor would, in some cultures anyway, be welcomed in an otherwise bland diet. In any event, there has been no established use, or adoption, of either type of material.

The projects did not fail because of the nutritional value of the protein

17. Pariser, Wallerstein, Corkery and Brown (1978), p. 87.
18. Ibid., pp. 83–5.

in FPCs. The few feeding trials with infants all gave good results, indicating that the quality of the protein was close to that of milk.[19] We will return later to Pariser's analysis of why these projects failed.

Oilseed Flours

The protein sources, other than fish, referred to in the FAO/WHO group's 1957 program of protein foods to be studied as substitutes for milk were all oilseed flours. Seeds are commonly classified as "oilseeds" if they contain at least 15% of fat (or oil) on a dry weight basis. Vegetable "oils" are, in chemical terms, fats, but the word "oil" is ordinarily used if they are liquid at room temperature. The protein of oilseeds typically provides about 20% of their calorie value. However, removal of the oil can leave materials containing 50% protein or more, so that they are indeed protein concentrates.

The pressing of seeds to obtain cooking oils has a long tradition. Olive oil, in particular, was a staple of the Greco-Roman world. In the tropics, peanuts and palm nuts, along with coconuts, have been important sources of cooking oil. Vegetable oils have also been in greater demand in the past 60 years with the development of chemical processing methods, including hydrogenation, that allow them to be used as substitutes for animal products such as butter and lard (shortening). Even with high-pressure equipment, the residue is left with 5% oil. Solvent extraction is more efficient, and most oilseed meals made in the technically advanced countries are the residues from extraction with hexane. This solvent is relatively inert, except to combustion, and there is no evidence that it has any harmful effects. In poor communities, where small operators have little capital, inefficient pressing methods may still be used, leaving residues that rapidly go rancid and can be used only as fertilizer or fuel.

By the 1950s the industrial processing of oilseeds was substantial, with the meals being used as economical protein supplements for farm animals. Even young chicks, which were thought of as delicate, made good use of them, and since they had even higher requirements for essential amino acids than infants, it seemed that the materials could be used as human foods also. There actually was already experience in the production of food-grade flours from some oilseeds, though they were used only at low levels, largely to improve the physical texture of foods such as bread rather than as nutritional supplements.

In general, the protein fraction of oilseeds is about 85% digestible, as compared with 90 to 95% for most animal foods. Their amino acid balance is also somewhat inferior, but they can still be valuable supplements in

19. Donoso, Maccioni and Monckeberg (1967).

human diets, with 3 parts of their protein (as a rough average) having the value of 2 parts of milk protein.[20] At the same time, oilseeds have certain advantages over raw materials such as fish: They are less perishable, so that processing can be spread out over the year, and they are grown all over the Third World in one form or another in great abundance.

Most oilseed flours contain unwelcome constituents, but ways have been developed to remove or inactivate most of them. One approach is to dissolve most of the protein by stirring the flour in an alkaline solution and then precipitating the protein from the extract by acidifying it. This kind of "protein isolate" has been used as a component of sophisticated processed foods in the United States, but the relatively high costs have made it irrelevant to helping poor people on a large scale. We will consider the work done with just three oilseeds, which have rather different characteristics.

PEANUTS

The peanut, or groundnut, *Arachis hypogaea,* is a native of South America but has been distributed throughout the tropical and subtropical world. It was estimated that in 1955 some 13 million tons of the nuts were harvested. It was also by far the major oilseed crop on the African continent, with much of it exported to Europe.[21]

The main advantage of peanuts is that they are very palatable. Peanut butter is made by shelling, roasting, skinning and then blending the nuts into a paste. Peanuts have also traditionally been used in China to prepare "milks," the nuts being ground to a paste, stirred with water and filtered.[22] The Indian government's Central Food Technology Research Institute set up a demonstration plant for producing such milk, but the idea was apparently not taken up commercially.[23] The same institute prepared batches of "Multi-Purpose Food" (MPF), a powder containing some 40% protein, mostly from peanut flour tested for freedom from aflatoxins (discussed later) and fortified with vitamins and minerals. A commercial food company also prepared batches, but only when government aid schemes or relief organizations purchased it; there was no regular commercial demand. In addition, the institute developed weaning food mixes, "Bal-Ahar" and "Bal-Amul" with 22% protein, of which peanut flour was an important constituent. These sold well to people with middle incomes and were also bought by organizations for emergency relief schemes, but production was not taken up commercially.[24]

20. Orr and Adair (1967).
21. Ibid., p. 13; Rosen (1958).
22. Rosen (1958), pp. 432–3.
23. Dean (1958), pp. 222–3.
24. Parpia (1969); Orr (1977), pp. 8–9.

In Nigeria, two commercial companies attempted, with government encouragement, to manufacture high-protein weaning foods with peanut flour as the base. The first of these, "Amama," was developed by the Glaxo Company as a cheaper alternative to its ordinary baby foods based on dried milk. In 1961, just as marketing was beginning, it was discovered that peanut meals could be seriously contaminated with carcinogens. Further research revealed that these "aflatoxins" came from fungi growing on improperly dried nuts after harvesting and that their presence could be measured with chemical tests.[25] In the meantime, however, the Glaxo Company, unwilling to be responsible for marketing a possibly dangerous product, closed its operation. In 1963 the Nigerian government undertook to purchase a second product, "Arlac," for free distribution through institutions and clinics. UNICEF also offered aid, on the understanding that the free distribution would be only a preliminary to the development of a regular commercial market for the product. The market did not materialize; UNICEF withdrew its support in 1968 and production ceased.[26]

Soybean Flour

The soybean (or soya bean) is native to China and has been an important food there for many centuries. From China it spread throughout East Asia and was brought to the United States from Japan in the 1850s. After a very slow start, it had become a major U.S. crop by the 1950s with an annual harvest of 14 million tons, which was a little more than one-half of the world harvest, and production has continued to increase. It is the world's major oilseed crop and well adapted to mechanized agriculture.[27]

The mature bean is almost spherical, small (less than 1 cm across), usually yellow and very hard. It contains about 20% oil and 40% protein. Most of the crop is processed by solvent extraction to yield oil that can be modified for cooking uses. The residual flour, which was originally of lesser economic value, has proved to be an excellent source of protein to supplement corn in pig and poultry rations. In particular, its protein is richer in lysine (the first-limiting amino acid in most grains) than is the protein of most other oilseeds. The protein in soy is rather low in methionine but, fortunately, synthetic DL-methionine is commercially available at an economic price and can be blended into compound animal feeds when necessary.[28]

One problem is that the beans contain at least two antinutrients – a trypsin inhibitor and a hemagglutinin, which depress the growth of exper-

25. Orr (1972), pp. 14, 16.
26. Ibid., p. 12.
27. Smith and Circle (1972).
28. Cravens and Sipos (1958).

imental animals. Fortunately, they are inactivated by heat. It is therefore standard practice for soy flour that is to be used for animal feeding to be heat-treated after the solvent-extraction step. The product is referred to commercially as "toasted soy flour" because it is somewhat browner than untoasted flour, though the heating is carried out under moist conditions.[29]

The highly successful use of soy as a supplementary protein at an economic price for pigs and poultry led to the recommendation, particularly by U.S. experts, that it be adopted as a supplementary protein source for infant feeding. Two of its proponents said, "Soybean protein outranks all of the other proposed supplements in the worldwide nutrition programs."[30] However, if peasant families in Africa, for example, with malnourished children are encouraged to grow soybeans for themselves, there is a problem. The mature seed, when cooked like other beans, by boiling in water, remains firm and somewhat bitter. And since Africans probably do not know any other way of preparing them, the crop will, almost certainly, be abandoned.

In Asia, where soy has been a traditional food, there are three main approaches to preparing palatable foods from them. One approach is to keep the seeds moist and in darkness for 4 days. They then sprout to a length of about 3 inches, and the sprouts can be eaten either raw or cooked.[31] They are similar to the mung bean sprouts more commonly sold in Western countries, but are not, of course, suitable as ingredients of weaning foods.

The second traditional treatment of soy, particularly in Indonesia, is to make *tempeh* by fermentation with a particular type of fungus, *Rhizopus*. The beans are soaked until the hulls can be easily removed by hand; they are then boiled for 30 minutes and spread out to dry. At this stage they are inoculated with the fungus, using a little of a previous batch of material, then patted into balls that can each be wrapped in a banana leaf. During a further 24 hours at room temperature, the fungal mycelium binds the beans together into a "cake" with a pleasant odor. It is then generally fried in slices and eaten within a day.[32] Some attempts were made to market a dried form of tempeh in Africa, but the product was not popular.[33]

The third type of processing begins with the preparation of a "milk." The beans are soaked in water for 3 hours or more and then ground, with the water, to a slurry, heated and held near the boiling point for about 20 minutes and finally filtered through cloth. The liquid, really a suspension, has to be drunk within the same day if no special precautions are taken to preserve it.[34] This type of material can be fortified with vitamins and min-

29. Liener (1980), pp. 40–9, 92–4.
30. Smith and Circle (1972), p. 19.
31. Ibid., p. 380.
32. Ibid., p. 404; Dean (1958), p. 218.
33. Orr and Adair (1967), p. 67.
34. Smith and Circle (1972), pp. 356–9.

erals. Soy milk has been sold commercially in Hong Kong ever since 1945 as "Vitasoy," being promoted as a thirst-quenching drink in competition with colas.[35] In this context, it is presumably consumed by older people with money to spend rather than weanlings at risk for malnutrition.

A government project in Indonesia undertook to spray-dry soy milk, which was sold as "Saridele," starting in 1957. The UN agencies gave assistance on the understanding that 30% of the output would be distributed free to charitable institutions. However, virtually all of the output was apparently sold on the retail market. Despite its fairly high price, it was popular among the fairly well off as a basis for making an after-dinner drink. After 10 years of operation the equipment needed replacing. The UN agencies were unwilling to provide further capital because the project did not appear to be achieving its original aim, and production ceased.[36]

In Asia soy milk is traditionally used mainly in the first step of *tofu* production. In this process calcium sulfate is added to the milk and a gelatinous curd formed. This is pressed gently in a series of small filter boxes, and each "cake" is then turned out. It still contains about 88% water, and only 6% protein, so that it is not a protein concentrate and has only a short "life" if not refrigerated. It has a bland flavor and can be either eaten as it is or deep-fried. About 65 to 75% of the protein originally present in the soybeans is recovered in tofu preparation.[37] Tofu is said to be an important protein source in a good proportion of East Asian households. However, it has not been developed as a supplementary protein source for malnourished infants in other areas.

Another product, "Pronutro," was specifically designed as a nutritional supplement. It is based on full-fat soy flour that has been heat-treated and given a special, additional treatment to remove the bitter flavor of soy. This was "developed, produced and marketed by a private food firm in South Africa without any governmental or international backing, as a high-protein foodstuff from vegetable protein sources."[38] Dr. H. de Muelenaere, a nutritionist associated with the development, has said that, when production began in 1962, the soybeans were imported, but the company indicated its interest in buying locally grown crops. By 1969 farmers had responded to the existence of this market by growing enough to meet the company's needs in full. The other major ingredient was locally grown corn. It was decided that the product should be "instant," that is, precooked so that it would appeal to overworked mothers of lower-income groups. It was also decided that it should be made attractive and first advertised to white groups. If

35. Orr (1972), p. 15.
36. Ibid., p. 13.
37. Orr and Adair (1967), pp. 67–8; Smith and Circle (1972), pp. 376–8.
38. De Muelenaere (1969), p. 266.

advertising and sales were directed solely, or mainly, to the black, Bantu people, they would suspect that it was a second-class food, lacking in prestige, even though clinical tests had demonstrated its value for malnourished black infants.[39]

The product proved attractive and economical to a wide range of people. After 1 year, it was being produced at a rate of 75 tons per month, and production was thought to have reached nearly 600 tons per month by 1967, but to have fallen back to 300 tons by 1976. By 1969 it was estimated that about one-half of that produced was going to Bantu customers. There are anecdotes to the effect that white customers found it so economical that, on the evidence of the advertising as to its high nutritional value, they began buying it for their pets; this in turn led some Bantu people to reject it on the grounds that they felt demeaned by the idea of sharing a food with dogs. The price is now said to have risen to a point where it is not within the reach of low-income groups, for whose children it was originally designed.

"Fortifex," another product based on soy flour, was marketed in Brazil in 1963. It was unpopular because of the "beany" flavor of the soy, and production ceased in 1966.[40] Many products based on soy have been discussed by reviewers, but none, apart from Pronutro, appears to have been commercially successful for any considerable period.[41]

Cottonseed Flour

Cotton plants are grown on a large scale because of the hairs (or lint) attached to the outside of the seeds. These can be separated and spun to form cotton thread and then woven into fabric. But for every ton of lint produced there remains, as a by-product, about 1.7 tons of seed. The worldwide yield of cottonseed is in the range of 15 to 20 million tons per year, and it contains about 18% oil and 20% crude protein. The United States has the biggest crop, but many Third World countries with malnutrition problems also grow sizable amounts.[42]

The seed, after the cotton lint has been removed, consists of about 45% by weight of hull (i.e., outer cortex) and 55% of kernel (or "meats"). The fibrous hull is of some value for ruminant animals, but is merely a source of undesirable bulk for single-stomach animals, especially if the seed is to be used in a weaning food for infants. With careful decortication there is no more than 5% fiber left in the final product. The oil is the most valuable part of the seed and, for technical reasons, is usually separated by screw-

39. Ibid., pp. 269–73; Orr (1972), p. 24.
40. Orr (1972), pp. 13–14.
41. Ibid., p. 66.
42. Altschul, Lyman and Thurber (1958), pp. 469–72.

pressing. The residual oil content of the meal can be below 6%, and this was made a requirement for cottonseed flours acceptable by PAG standards. The crude protein level in such meal is approximately 50%.[43]

Cottonseed contains a pigment, gossypol, which has adverse effects on animals. However, its molecule has reactive chemical groups (both aldehyde and diphenyl groups). Gossypol is stable when in the intact seed, where it is stored in pigment glands, but during screw-pressing the glands burst, and at the elevated temperature in the press the gossypol reacts with the protein and probably with other components in the seed. This "bound gossypol" becomes indigestible and apparently harmless. Cottonseed varieties have been bred without pigment glands, but these were not commercially grown when high-protein products based on cottonseed flour were being developed in the 1950s and 1960s. Amino acid analysis showed a somewhat lower level of lysine in cottonseed meal than in soy, but a slightly higher methionine level.[44]

In Central America during that period, cottonseed was the major oilseed, and not fully utilized. The Institute for Nutrition in Central America and Panama (INCAP) therefore set out to develop mixtures based on cottonseed flour as the major protein source, with a high nutritive value. INCAP's mixture 9B consisted of 28 parts of ground corn, 28 of sorghum grain, 38 of cottonseed flour, 3 of dried yeast, 1 of calcium carbonate plus synthetic vitamin A. This gave good results in preliminary laboratory tests with chicks and rats. Then it was tested, in a cooked form, with young children recovering in the hospital from malnutrition, and it performed nearly as well as mixtures based on milk powder. Finally, it was tested for acceptability among needy families in rural areas as a substitute for the cereals they were accustomed to cooking into an *atole,* or thin gruel, with added flavorings, for their young children.[45]

Given the satisfactory results of these tests, INCAP decided to encourage commercial production of this mixture (and others of the same general type), under license and supervision from INCAP, using the trade name "Incaparina" and with the price fixed as low as possible. It was retailed as 75-g portions in plastic bags, each enough for three glasses of atole, and also sold in bulk to charitable institutions. Commercial production of Incaparina began in Guatemala in 1961 and was still in production in 1980 at the rate of 300 tons per month. In other Central American countries, production was started but ceased after a few years. There were various reasons for this, including competition from material donated by international agen-

43. Ibid., pp. 484–505; Protein Advisory Group (1970).
44. Altschul, Lyman and Thurber (1958), pp. 489–92, 500.
45. Scrimshaw et al. (1961); INCAP (1962).

cies.[46] Incaparina was produced in Colombia for 12 years, from 1963 to 1975, by the Quaker Oats Company. In 1969 it appeared to be doing well and was sold at a price (based on equivalent quantities of protein) roughly one-eighth that of milk powder, but higher ingredient prices and increasing competition from other weaning foods finally made its production uneconomical.[47]

There has been vigorous debate over whether Incaparina should be regarded as a success or a failure. One paper pointed out that the Guatemalan production in relation to a population of about 5.6 million corresponded to only 1.2 lb. per person per year. There was also evidence that the poor bought less than the better off. From survey data on the use of Incaparina, it was estimated that 87% of the children got none, 5% consumed between one-fifth and one-tenth of the recommended level and 6% consumed a little more than one-third.[48] The director of INCAP wrote in 1976 that recent increases in the prices of ingredients had resulted in the price of Incaparina being placed "a little beyond the purchasing power of the people who needed it most – the underprivileged."[49]

Scrimshaw, the director of INCAP when Incaparina was being developed, has defended the record:

> There was never any claim or expectation that a commercially produced and marketed vegetable mixture could "solve" the problem of protein–calorie malnutrition in Guatemala or any other country. This could only be realized through an increase in income among the groups in need, or by effective programs of subsidized distribution. There was, however, a need for a weaning food that could be recommended for malnourished mothers and children in a country where milk was too costly and in short supply... it would be the height of arrogance to deny developing countries the benefits of a nutritionally comparable food at the lowest practical cost. The objective, then, was to provide a beverage with the nutritional equivalent of milk ... for the benefit of a sector of the population with modest purchasing power, leaving to other programs the problem of reaching that part of the population unable to purchase weaning foods.[50]

Single-Cell Proteins

Single-cell proteins (SCPs) include protein concentrates prepared from any type of single-celled micro-organism. There are several readily available

46. Scrimshaw (1980); Shaw (1969), pp. 321–2; Orr (1972), pp. 7–11; Orr (1977), p. 4.
47. Dimino (1969).
48. Wise (1980).
49. Icaza (1976), p. 37.
50. Scrimshaw (1980), p. 1.

books on this interesting group of materials.[51] One attraction of growing micro-organisms on a large scale as a food supplement is that they can reproduce so rapidly. One proponent of their use has compared the daily production of 0.5 kg milk protein by a 500-kg cow to the 1,250 kg protein that could (in theory) be produced by 24 hours' fermentation of the same weight of yeast.[52]

Yeasts had been grown on carbohydrate wastes for many years, but French work in the 1950s demonstrated that some strains of yeast could also use hydrocarbons as their energy source, and "protein from petroleum" became a catchy concept when there was a general fear of a world shortage of protein from conventional sources.[53] Production commenced, at least on a pilot-plant scale, of a variety of organisms, including bacteria and fungi as well as yeasts, and in general the products gave good results in animal feeding tests, with both rats and pigs. However, there were considerable difficulties in producing materials that satisfied concerns about their safety as human food ingredients.

The first problem was that, because micro-organisms reproduce so rapidly a larger portion of their nitrogen is present in nucleic acids (i.e., their genetic material) than is the case with conventional plant or animal foods. Two constituents of the genetic material are purines, adenine and guanine, and after the digestion and absorption of foods, these components are largely oxidized to uric acid. In most mammals an enzyme then converts the uric acid to another more soluble compound, allantoin, which is freely excreted by the kidneys. However, the primates, including humans, do not have this enzyme, and high concentrations of uric acid can crystallize out, as stones both in the urinary system and in joints, causing the pains of gout.

There is considerable variability among individuals as to the quantity of nucleic acids that they can tolerate, and safety limits have to be set for the most sensitive. The general recommendation has been that SCPs should contribute no more than 2 g nucleic acids to a day's diet. Since the true protein content of yeast, for example, may be no more than four times the nucleic acid content, such a product cannot therefore be safely consumed at a level contributing more than 8 g protein per day.[54] Other yeasts have a more favorable ratio, but some bacterial products have a ratio of little more than 3:1.[55] It is technically possible to reduce the nucleic acid content of products by further processing, but of course this adds significantly to the costs of production.

51. E.g., Mateles and Tannenbaum (1968); Tannenbaum and Wang (1975); Moo-Young and Gregory (1986).
52. Champagnat, Vernet, Lainé and Filosa (1963), p. 14.
53. Ibid., p. 13.
54. Moo-Young and Gregory (1986), pp. 109–11.
55. Mauron (1980–1), pp. 73–4.

The second problem with a number of SCP products has been that, although they have passed exacting toxicological tests with a variety of animal species, about 15% of the human subjects receiving yeasts developed eczema and other allergic reactions, and bacteria have caused vomiting and diarrhea.[56]

A third concern has been that production facilities might be contaminated by less desirable micro-organisms without their being detected. Precautions undertaken to minimize this risk have included the sterilization of very large air (or oxygen) intakes, which again has raised costs significantly.[57]

Because of these considerations, new yeast and bacterial products have so far been used only as protein supplements for animal feeding, though one fungal product has been marketed as a food.[58]

Lysine-Enriched Grain

The protein of grains was known to be first-limiting in lysine, as judged by feeding experiments with rats and by comparison with whole-egg protein. Plant breeders at Purdue University therefore set out to produce corn (maize) of improved lysine content. In 1964 they published a preliminary description of "opaque-2 corn" with a lysine content of approximately 4 g per 100 g crude protein, as compared with 2.5 to 3 g per 100 g in ordinary corn, and nearly a doubling of the tryptophan level.[59] In 1966 a large conference was held to present the first nutritional evaluations of the new corn. The introductory speaker suggested that it would become a historic meeting and that "within the next five years millions of undernourished people in Latin America should find their diets improved markedly due to the availability of high-lysine corn."[60] Feeding experiments with young animals confirmed its value, but farmers were not interested in growing the new variety, since the yields were relatively low and its soft kernels made it exceptionally subject to insect attack and unsuitable for the usual milling equipment. A further 20 years of work resulted in "protein quality maize" without these disadvantages, though some agronomic questions remained, and it is still not in general use.[61]

A further approach was to fortify ordinary staple grain foods with synthetic lysine. In 1956 a scientist working for the DuPont Chemical Company, which had begun to manufacture lysine, urged that it would be a beneficial additive for human diets, even in the United States, for the young and the

56. Waslien, Calloway and Margen (1969); Scrimshaw and Dillon (1979).
57. Moo-Young and Gregory (1986), pp. 97–102.
58. Ibid., pp. 19–26.
59. Mertz, Bates and Nelson (1964).
60. Liebenow (1966).
61. Mertz (1992).

disadvantaged.[62] In a number of trials, young Third World children recovering from protein–calorie malnutrition in medical units retained more nitrogen when their mainly cereal diets were supplemented with lysine.[63]

This result encouraged a U.S. government agency to finance large projects in Guatemala, Tunisia and Thailand to study the value of fortifying corn, wheat and rice, respectively, with lysine.[64] In Thailand the rice was also supplemented with threonine. In each country different groups received either no supplementation, vitamin and mineral additions only, or amino acid(s) plus vitamins and minerals. The areas studied were chosen because grains formed a particularly large proportion of the local diets. After a great deal of work, the project leaders summed up their results: "In Guatemala there was very slight and inconclusive evidence of a decrease in morbidity and mortality rates; the other studies showed no detectable benefits."[65] It was concluded that low energy intake was the factor most responsible for restricting children's growth in these areas, along with possible marginal deficiency of zinc and/or other trace minerals. Under such conditions a response to lysine was not to be expected.[66]

Summing Up

From 1956 on, the Protein Advisory Group encouraged the development of many new types of high-protein material. Funding was provided by individual governments, and in 1970 the United Nations agreed on the need for a $75 million program.

The first general goal was to upgrade the fish meals and oilseed meals already used in animal feeding to a quality suitable for use with young children. But this led to a variety of problems. One was that to produce fish protein concentrate at a low cost required whole fish to be used, and their intestinal contents, even after cooking, were technically unhygienic. Another problem was the undesirably high content of fluorine in their bones.

The West African program to market an infant formula based on peanut flour came to an end in 1961 because of the discovery that peanuts stored in humid tropical conditions commonly became contaminated with a carcinogenic fungus. Preparations based on "milks" made from soy were developed in Asia, but were generally too expensive for adoption as a low-cost weaning food. A precooked mix, Pronutro, with soy flour as its main protein source was marketed by a commercial company in South Africa. It

62. Flodin (1956).
63. Austin (1979), pp. 269–81.
64. Ibid., pp. 131–268.
65. Kennedy, El Lozy and Gershoff (1979), pp. 27–8.
66. Vaghefi, Makdani and Mickelsen (1974).

proved palatable and sales reached 600 tons per month, but increases in the prices of raw materials then put it out of reach of those most in need.

The Institute for Nutrition in Central America and Panama developed mixes with cottonseed flour as the high-protein component. These were cooked and used as gruels for young children. They were marketed in several Central American countries but, again, only a small proportion was consumed by the poorest groups.

There were many attempts to produce single-cell proteins (yeasts, fungi or bacteria) as a supplementary food for children, but problems of quality control and adverse reactions to them among some people prevented production and commercial marketing on a significant scale.

These and other problems with high-lysine grains were great disappointments to those involved. Even the costs of simple packaging and transport could put otherwise economic materials beyond the purchasing power of villagers. It had been the original policy to have everything produced in the area of need. Yet to quote someone concerned with the U.S. fish protein project: "FPC represented a technological application conceived in the industrialized world in an environment and time period that were relatively free from constraints on capital, resources and expertise. This product was to be applied in the developing world where all of these constraints existed in abundance."[67]

With the altered interpretation of the malnutrition problem that will be described in the following chapter, most of these developments lost their funding, so that we do not know what they might, or might not, have achieved.

67. Pariser, Wallerstein, Corkery and Brown (1978), p. 231.

10 Reappraisals of the Third World Problem, 1955–1990

In chapter 9 we saw some of the practical problems that arose in trying to provide additional protein sources for people in the poorer countries of the world. In the same period that these attempts were being made, there was a progressive reassessment of the human requirements for protein and other nutrients that was to alter greatly the general view of the "world protein crisis" and the cause of the major malnutrition problems among young children in the Third World. This forms the subject of the present chapter, which picks up where we left off at the end of Chapter 8.

Bulk and Limiting Calories

From 1955 on, there was an increasing interest in the calorie component of infant malnutrition, in kwashiorkor as well as marasmus. Dr. Hebe Welbourn, a medical officer with a long experience with child welfare clinics in Uganda, wrote that the eating behavior of the children she saw had also been well described by Williams for West African children: "Each one pinches off a small piece [of maize porridge] from the lump. He gently shapes it with his fingers, dips it in the soup, and puts it in his mouth. He delicately licks his fingers and repeats the operation. The whole process is deliberate and mature; there is no scramble or hurry. But their appetites are generally poor."[1] Welbourn, returning to her own experience, wrote:

> One of the most noticeable features of all cases of kwashiorkor was anorexia.... The daily caloric intake recommended by the League of Nations (1936) for children aged 2–3 years was 1,000 kcalories. Therefore most of the children, but especially those of peasant families [for which we recorded a mean intake of 762 kcal,] were taking [in] insufficient [amounts]. This was, no doubt, partly due to their poor appetites, and partly due to having only two main meals daily. A child's stomach is small, and his needs are relatively large, while African food is very bulky. He

1. Williams (1938), pp. 97–102.

cannot pack in enough of this bulky food in two meals to supply his caloric needs for the day.[2]

In her summary, she stated:

It has been shown that the diets of Baganda children are inadequate from the time that their mothers' milk alone is no longer sufficient for their growing needs. The deficiency in diet which is maximal during the weaning period is associated with a profound disturbance of growth and the appearance of signs of kwashiorkor.... Severe kwashiorkor, as seen in hospital, would appear to be the end result of a prolonged period of feeding on an unbalanced diet which is mainly of carbohydrate and is deficient in protein. The protein deficiency is usually aggravated by the occurrence of infections, particularly those of the upper respiratory tract, which are often followed by prolonged anorexia.[3]

In 1953 Dr. Katerina Rhodes had presented data at the Macy Foundation's conference in Jamaica indicating that the kwashiorkor victims she had studied there had had a low calorie intake while the disease was developing.[4] This was contrary to the previous assumption that such "sugar babies" consumed ample or even excessive calories, and her data were closely questioned. However, in 1957 she published them in full and concluded that, although the carbohydrate intake was high in relation to protein, there was still an absolute calorie deficit. She also referred to similar data from India, Indonesia and South Africa.[5]

In 1958 a group working in South Africa described the results of a trial in which 60 kwashiorkor patients were treated with fresh skim milk supplemented with 5% of either carbohydrate (dextrimaltose) or protein (casein) and, for a third treatment, dried skim milk alone (15 g per 100 g water). The recovery rate and the rise in serum albumin levels were similar in all the groups. It was concluded that expensive protein supplements to skim milk were not necessary, since the addition of carbohydrate as a source of additional calories was equally effective.[6]

In the following year, D. B. Jelliffe, who had worked in a number of Third World countries with kwashiorkor and marasmus patients, wrote an extensive review, introducing the term "protein–calorie malnutrition" in its title.[7] This quickly came into general use, with "PCM" as the accepted abbreviation after the FAO/WHO Expert Committee adopted it at its meeting in 1961. The committee report stated:

2. Welbourn (1955), p. 105.
3. Ibid., p. 169.
4. Waterlow (1955), pp. 43–5.
5. Rhodes (1957), pp. 143–50, 166–7.
6. Pretorius and Smit (1958).
7. Jelliffe (1959).

In developing countries a high percentage of children from lower-income families may show retardation of growth and development due to inadequate feeding. The prevalence of protein–calorie-deficiency diseases is increasing in some countries as a consequence of the rapid spread of industrialization and urbanization and the resulting adverse changes in food habits. One common result is the premature termination of breast-feeding and the use of overdiluted cow's milk, thin gruels and cooking water from cereals as substitutes.... As the widespread use of the term *marasmic kwashiorkor* suggests, kwashiorkor may be superimposed on any degree of marasmus. The high prevalence of deficiency-disease syndromes makes it urgent that attention be directed to practical methods for their prevention. Kwashiorkor has tended to engage the exclusive attention of many workers. The attention of these investigators and of those responsible for preventive and corrective programmes should be directed, without decreasing the interest in kwashiorkor, to all aspects of the problem of protein–calorie-deficiency disease.[8]

Recommended Protein Levels for Infants

In 1958 the U.S. Food and Nutrition Board issued its regular four-yearly revision of recommended dietary allowances (RDA). For the first time, there was a blank space in its tables – the RDA for protein among infants in the first year of life. An asterisk indicated that breast feeding was recommended for this period and that it provided no more than 2 g protein per kilogram body weight daily. The majority of formulas had a higher level of protein, which provided approximately 3.5 g/kg, and these, too, had proved to be successful in practice. Opinion had been divided as to what value should be listed in the table.[9] Reviewers in the following year set out the points of debate in more detail. They argued that the RDAs were designed for their own country (the United States), which had an abundance of food. It had been stated that the intention of the RDAs was to provide "ample allowances" and that "they may be more generous than would be practical for feeding large groups under conditions of limited food supply or economic stringency."[10] Moreover, it could not be assumed that human infants were able to use cow's milk protein as efficiently as breast milk protein. "The protein of unheated cow's milk was less well digested because of the hard curd formed in the infant stomach, but there was not the same problem with heated or dried milk preparations."[11]

Waterlow and his colleagues in Jamaica reported, in 1961, the results of a series of balance studies in children (12 months old on average) who were being treated for kwashiorkor. Surprisingly in view of the pancreatic damage

8. Joint FAO/WHO Expert Committee on Nutrition (1962), pp. 23–4.
9. Food and Nutrition Board (1958).
10. Gordon and Ganzon (1959), p. 503.
11. Ibid., p. 508.

associated with the disease, the cow's milk protein that the children received was digested with 90% efficiency. The secretion of protein-digesting enzymes into the gut was therefore adequate, though it may have had less than the usual margin of safety. Also, they found that even 2 g protein per kilogram body weight per day supported a rate of nitrogen retention three times that of healthy babies. The "catch-up" phenomenon was well known, but it seemed reasonable to conclude that, if this was possible on 2 g cow's milk protein per kilogram, then the normal rate of growth should easily be achieved with the proposed standard of 1.5 g/kg for healthy babies.[12] Waterlow also found that the rate of recovery of babies depended very much on their energy intakes. He obtained improved responses by giving 145 kcal/kg daily in mixtures with 10.5 PCals%. He suggested that, in the past, the rate of recovery from kwashiorkor had most commonly been limited by the number of calories provided rather than the amount of protein.[13]

Supplementary Calories Needed

In 1962 B. M. Nicol, who had become Autret's deputy at FAO, reviewed "the utilization of protein-rich foods in the prevention of protein–calorie deficiency diseases." He wrote:

> Probably the commonest cause of protein–calorie malnutrition is lack of food, which may result from... an overall shortage of supplies affecting the whole community, but more often results from the parents' failure to realize that nutritional requirements of young children are relatively much greater than their own. [Let us take as an example]... the diet of a poor Indonesian community. It contained rice, a little fish, sweet potatoes, vegetable oil, fruit, green vegetables and seasoning. This diet provides adequate energy and protein if enough of it is available for consumption. The amounts actually fed to weaned infants and pre-school children only provide two-thirds of their energy and protein needs. Small amounts of dried skimmed milk or full-fat soya flour correct the protein deficiency but do not supply sufficient calories to bridge the gap between intake and requirements which must be achieved for optimal protein utilization. For this purpose the supplement should also provide concentrated sources of energy such as sugar or vegetable oil. The diets eaten in Southern Nigeria are very bulky, being composed of starchy roots, very small amounts of meat and pulses, and reasonable quantities of green vegetables. These staple foods are freely available but even adults cannot consume sufficient of this diet to meet their needs. Pre-school children can eat only enough to provide 65% and 70% of their requirements for protein and calories respectively. In view of its physical bulk, supplementation of such a diet is undesirable. The replacement of roots by a good quality cereal such as rice, to provide a more concentrated source of protein and calories, is necessary.[14]

12. Waterlow and Wills (1960), pp. 194, 196.
13. Waterlow (1961), pp. 21–2.
14. Nicol (1962), pp. 152–5.

It is interesting that this view was expressed by a senior FAO official during the period when the PAG was accelerating its work on new types of protein concentrates.

In 1963 Graham and his colleagues, working in Peru, said that use of the term "protein–calorie malnutrition" for the whole spectrum of cases from kwashiorkor to marasmus was resulting in unsuitable treatment of the latter condition, which was much the commoner. Children with marasmus had a serious caloric deficit and "failure to appreciate [this] leads to a waste of expensive protein and to continued failure in treatment."[15] In a second paper they reported that 2.0 g protein per kilogram body weight daily was sufficient for hospitalized malnourished children but that daily caloric intakes as high as 175 kcal/kg were highly beneficial to rapid recovery.[16]

In 1965 FAO/WHO published its second report entitled "Protein Requirements." The difference between the approach used by the Expert Group on this occasion and that described in the 1957 report will be considered later in a comparison of all the reports on the subject. The main point to be noted here is that the estimates of requirement were much reduced. For example, the need of a 1-year-old child for protein of the quality of breast milk was estimated to be 1.1 to 1.2 g/kg per day, as contrasted with 2.0 g/kg per day specified in the previous report.[17]

In 1966 Chan and Waterlow reported that 1.25 g cow's milk protein per kilogram received daily seemed to give optimum growth in children who had almost recovered from PCM when they were receiving a total of 120 total kcal/kg daily. The balance of the calories came from sucrose, corn starch and peanut oil.[18] This mixture had a protein concentration of only 4.2 PCals% (i.e., lower than that of breast milk).

McLaren wrote that, though the incidence of marasmus was increasing at a faster rate than that of kwashiorkor and was a more important type of PCM, it was neglected.[19] "Millions of dollars and years of effort that have gone into developing these [high-protein] foods would have been better spent on efforts to preserve the practice of breast feeding... being abandoned everywhere."[20] In 1967 Jelliffe wrote that, because tubers and plantains were so bulky owing to their water and fiber content, they needed partial replacement with "compact calories" and that red palm oil, sugar and avocados, for example, formed a good mashable source of them.[21]

It would be wrong to give the impression that the papers just quoted were

15. Graham and Morales (1963), p. 486.
16. Graham, Cordano and Baertl (1963).
17. Joint FAO/WHO Expert Group (1965).
18. Chan and Waterlow (1966).
19. McLaren (1966).
20. McLaren (1967).
21. Jelliffe (1967).

typical of the period 1955–67. They were not. A much larger number of papers dealt almost exclusively with the evaluation of different kinds of protein mixtures for children and gave little or no attention to the question of the caloric adequacy of the whole diets in which these high-protein foods might play a part. Nevertheless, it is interesting that experienced people were already trying to draw attention to the energy aspect at a time when the UN system seemed to be entirely concerned with the "world protein crisis." The UN paper "International Action to Avert the Impending Protein Crisis" was approved in 1968, and governments were asked to report what actions they were taking on the "protein food problem."[22]

Marasmus and Kwashiorkor

The seventh meeting of the FAO/WHO Expert Committee on Nutrition accepted, with only a few minor reservations, the 1965 report on protein requirements, but did not discuss the implications of the large reduction in the estimates. In connection with reports of the increasing incidence of marasmus as a result of urbanization, they commented that the PAG's protein-rich food program was "designed for the prevention of any form of protein–calorie deficiency."[23] This statement may have been made in direct response to the comments of Graham and others (quoted earlier), who were of exactly the opposite opinion.

Some further differences of opinion emerged at a conference organized by McCance and Widdowson in England in 1967. McCance, who had by then spent some years at the Medical Research Council's unit in Uganda, opened the discussion by saying:

> The pioneer workers were right after all, and there really are two quite separate syndromes ... *marasmus*, which we can describe as an insufficiency of all food, however good, and *kwashiorkor*, which is essentially a deficiency of protein in a diet potentially high in calories.... Both prevent growth, marasmus for obvious reasons which have been recognized for centuries, and kwashiorkor because protein deficiency produces anorexia and ultimately destroys all desire to eat.

He also argued that "protein–calorie deficiencies" was a better term than "protein–calorie malnutrition," which implied a single syndrome.[24] Hansen, in contrast, speaking from his experience in South Africa, said that, in every type of PCM,

> a deficiency of protein is the overriding problem, which may or may not be complicated by a calorie deficiency.... From the clinical and public health point of view

22. Kertesz (1968), p. 72.
23. Joint FAO/WHO Expert Committee on Nutrition (1967), pp. 8–10, 60.
24. McCance (1968), pp. 1–2.

the similarities [between kwashiorkor and marasmus] are far more telling than the differences in detail.... Both are associated with a protein-deficient diet.... Individual children can drift from one syndrome to another.[25]

At the same meeting Gopalan described the studies carried out by workers in India and their surprise at being unable to find any difference between the diets of children who developed kwashiorkor and of those who developed marasmus. There was no evidence of any of the former having been "force-fed" starchy foods. The average daily diet of these children contributed 1.8 g protein and 85 kcal/kg body wt. Even if the protein mixture, being mostly vegetable, had a value only 60% that of "reference protein," this would still be equivalent to 1.0 g reference protein per kilogram and so would apparently meet the requirement, whereas the caloric intake was clearly inadequate. He hypothesized that marasmus represented an extreme form of *adaptation* to such a protein–calorie deficiency, whereas kwashiorkor represented a *dysadaptation* – perhaps as a result of some further stress such as infection. Possibly the different syndromes involved differences in endocrine response to the stresses on the children.[26]

McLaren drew attention to PCM being so often complicated by vitamin A deficiency, which had not received proper attention.[27] Other participants referred to the potassium and magnesium status of kwashiorkor patients as a cause of concern and to the need for appropriate supplements to minimize mortality in the most seriously ill.[28] These comments emphasized the danger of considering only one or two nutrients at a time.

From this point on, the relevant papers and discussions centered around three themes, which we will follow separately for the remainder of this chapter:

1. the idea that kwashiorkor represented a dysadaptation,
2. successive estimates of human requirements for protein and
3. the controversy as to whether it was primarily protein deficiency that caused protein–calorie malnutrition.

Kwashiorkor as the Result of Dysadaptation

Marasmus, sometimes considered to be simple "wasting," is not usually accompanied by a serious fall in the level of serum albumin. Nor does the liver become fatty. In kwashiorkor both of these changes occur, even though there is less wasting than in marasmus. It was difficult to understand why,

25. Hansen (1968), p. 46.
26. Gopalan (1968), pp. 52–3, 57–8; Jaya Rao, Srikantia and Gopalan (1968).
27. Gopalan (1967); McLaren (1968).
28. Wharton, Howells and McCance (1967); Wharton (1968), pp. 150–1; Garrow, Smith and Ward (1968), pp. 61–87; Whitehead (1969), p. 41.

in marasmus, muscle protein was catabolized to provide the liver with the material needed to synthesize the different serum proteins so crucial for preventing edema, while in kwashiorkor this did not happen, at any rate to the same extent.

As discussed by Whitehead and Alleyne in 1972, on the "classical" theory that kwashiorkor developed when carbohydrate intake was high or normal, the endocrine response (e.g., raised serum insulin levels) would inhibit muscle protein breakdown, while the virtual starvation of marasmus (with e.g., lower levels of insulin) would encourage such breakdown and so add to the bloodstream a complete mixture of free amino acids, which could be used by the liver for resynthesis of serum proteins.[29] It was accepted that, once kwashiorkor had developed, one of the effects was a loss of appetite, but by this time the damage had presumably been done.

Cortisol (hydrocortisone), an important hormone secreted by the adrenal gland, has an activity generally opposite to that of insulin. In particular, it increases protein breakdown in peripheral tissues, but stimulates protein synthesis in the liver. Stress stimulates its secretion into the bloodstream. In 1973 Whitehead's group reported results from a detailed study of the levels of cortisol and other hormones in children developing PCM in Uganda, as well as a review of the literature. The average serum cortisol levels of children with severe kwashiorkor were raised, but there were some individual children with normal values. The values showed a close correlation with the severity of the children's recent infections. Also, there was a similar average elevation in cases of marasmus. Values were normal among children in the early stages of kwashiorkor, when serum albumin values were only slightly reduced. Analyses for serum insulin levels after fasting overnight indicated that values were raised in the early stages and then, with fully developed kwashiorkor, fell to subnormal levels. The decreases again seemed to coincide with bouts of infection. Finally, a rise in the level of growth hormone in the serum was found, but only in combination with severe reduction in serum albumin levels, that is, in the most serious cases. To sum up, it appeared that at first the only abnormality was a raised insulin level but that, later, this level fell, and cortisol and growth hormone levels were raised. The increase in growth hormone concentration was seen in kwashiorkor but not at all in marasmus and was thought to be particularly significant.[30]

It had been the presumption of the group in Uganda that dietary differences were primarily responsible for the development of two different diseases. However, Jaya Rao from India presented another hypothesis based on the assumption that there were no significant differences in diet between

29. Whitehead and Alleyne (1972), pp. 73, 76–7.
30. Lunn, Whitehead, Hay and Baker (1973).

those developing kwashiorkor and those suffering from marasmus. Most children were able to respond to the stress of a diet extremely low in calories and low (or marginal) in protein, to adapt and gradually become marasmic. They did this, primarily, by greatly raising the level of circulating cortisol – which stimulated the movement of amino acids from muscle to liver. Children who did not react in this way to produce a sufficient increase in cortisol suffered the effects of a liver starved of the limiting amino acids needed to synthesize serum proteins.[31] Presumably the failure of the adrenal cortex to respond was due either to some additional stress, such as infection, or to an inherent, genetic characteristic of the individual.

This hypothesis seems not to have been generally accepted. Two replies appeared in the *Lancet,* making a number of points. First, it had been found elsewhere that cortisol levels were raised only in cases of PCM where there was infection and that the rise then seemed independent of whether the basic condition was one of kwashiorkor or marasmus. Second, the proportional incidence of the two diseases varied from one area to another, and this could be related to the prevalent dietary pattern. Finally, the hypothesis ignored the finding that a consistent, kwashiorkor-like state could be produced in animals purely by dietary manipulation.[32]

Because, once a disease is established, there may be complicating effects of anorexia, Whitehead and co-workers carried out a long-term study of children *in their homes* for the first 3 years of their life in rural communities in both the Gambia (West Africa) and Uganda. It appeared that children in Uganda, where kwashiorkor was more common than marasmus, were better off in the first year of life, then suffered a greater shortage of food in the second. However, they did not appear to react to this, as would be expected, with lower insulin and higher cortisol levels in their blood. In fact, it was the children in the Gambia, where marasmus was more common, who showed this response, perhaps because of the greater incidence of infections, mostly gastrointestinal. It is interesting that, although the mean level of protein in the diets of the two groups of children was very similar, at about 8.4% total calories, there was greater variability in the level from one child to another in Uganda. And among the individuals whose food intake was studied, it was those with the lowest protein intake who seemed most at risk from malnutrition.[33] The picture was complicated and provided some support both for those who argued that kwashiorkor represented a failure of endocrine adaptation (since cortisol levels were lower in Uganda) and for the opposing view that it was due to protein deficiency. Most surprisingly, the data did not fit what seemed to be the noncontroversial

31. Jaya Rao (1974).
32. Rosen, Buchanan and Hansen (1974); Waterlow (1974).
33. Whitehead, Coward, Lunn and Rutishauser (1977).

Table 10.1. *Calculations leading to estimates for safe protein allowances in successive FAO/WHO reports*

Year of report	Mean daily requirement for good protein (g/kg body wt)		Allowance for individual variability (%)	Approximate protein score for poorest national diet		Safe protein allowance on poorest diet (g/kg body wt)	
	A	B		A	B	A	B
1957	0.35	1.30	+50	60	60	0.88	3.25
1965	0.59	0.92	+20	60	60	1.18	1.84
1973	0.44	0.98	+30	70	70	0.82	1.82
1985	0.60	1.08	+25[a] +27.5[c]	85[b]	51[b]	0.88	2.70

Note: Columns labeled A contain values calculated for young adult males; columns labeled B contain those for 1- to 2-year-old children.
[a]For young adult males.
[b]The actual scores, based on amino acid analyses, were 100 and 58, respectively, but an additional 85% digestibility factor was applied to the poorest diets with high proportion of vegetable protein.
[c]For 1- to 2-year-old children.

view that marasmus could be simply explained by inadequate food intake, since the Gambian children ate more in their second year in relation to their size.

What Are Normal Protein Requirements?

In this section, we will review the series of four FAO/WHO reports on this subject published in 1957, 1965, 1973 and 1985, and refer to other reports only when they use different means to arrive at their estimates. We will also concentrate on just two classes of subjects – 1-year-old children and young adult males – the former because they are the ones at risk of developing kwashiorkor in the Third World, and the latter because they are the group that has been most used for direct experimentation. All the reports followed the same general procedure, that is to say, they first considered the requirement for proteins of excellent quality (i.e., those of milk or eggs) and then tried to assess how the requirement should be adjusted upward for other protein sources. The results are summarized in Table 10.1.

Because so much of the work in nutrition in the previous 40 years had been concerned with vitamins, and so little with protein, the authors of the first (1957) report had to go back to Sherman's paper, published in 1920, for their estimate of adults' daily requirements.[34] They assumed a value of 0.35 g/kg body wt for good-quality protein (that of eggs or milk). Their second assumption was that young suckling infants needed all of the protein (2 g/kg) they received from breast milk.[35] Scrimshaw has referred to the debates within the committee as to the rate at which the requirement declined as a child aged, but the final estimate for a 1-year-old child was that, on average, he or she required 1.3 g/kg.[36]

Next came the problem of converting these values to safe practical allowances. As seen in Table 10.1, the decision was made first to increase the values by 50% to allow for individual variability – that is, for some children needing more than the average. Then there was the problem of allowing for the lower quality of protein in practical diets. This was based on chemical scoring similar to that developed by Mitchell and others, as discussed in Chapter 7. From an examination of dietary patterns in different countries it appeared that the worst combination likely to be encountered had a value approximately 60% that of ideal protein. From the inverse of this percentage it followed that 1.67 g of such protein was equivalent to 1 g of the ideal. Therefore, as seen in the table, a 1-year-old's safe daily allowance in such circumstances would be $1.3 \times 1.5 \times 1.67 = 3.25$ g/kg body wt.

The 1965 report followed a different approach. It reviewed the data available as to how much nitrogen per day was lost by human subjects when they were on a diet free of protein (and nitrogen), through feces, urine, skin, sweat and so on. It was argued that the "perfect" dietary protein, providing the same quantity of nitrogen, would exactly replace these "obligatory" losses and bring the subject back into balance. Protein needed for body growth, growth of a fetus or milk secretion would also be obtained with 100% efficiency from such a perfect dietary protein.[37] However, the estimates were increased by 10% to make allowance for the effect that stresses such as minor infections and trauma were believed to have on protein breakdown and losses from the body.[38] The requirements of ideal protein, before allowance for individual variability, came out to 0.92 g/kg for 1-year-olds and 0.59 g/kg for adult men. Practical recommendations were made by multiplying by a small variability factor and a "scoring" factor.

The next study, published in 1973, considered energy as well as protein. Since the preceding study, there had been a number of tests of its assumption that egg protein could compensate for obligatory losses with virtually 100% efficiency. In practice, with young adults, the finding was that it could not. The recalculated obligatory nitrogen losses (incorporating some new data)

34. Sherman (1920); Irwin and Hegsted (1971a).
35. FAO Committee on Protein Requirements (1957), p. 15.
36. Scrimshaw (1976), p. 138.
37. Joint FAO/WHO Expert Group (1965), pp. 10–16.
38. Ibid., pp. 17–18.

were lower at 54 mg/kg per day, equivalent to 338 mg crude protein, but 463 mg egg protein had been found necessary, on average, to achieve balance. In addition, this was very similar to the quantities required to obtain nitrogen balance with soy flour or any of a variety of mixtures of plant and animal proteins.[39]

Two conclusions were drawn. One was that supplying dietary protein increased the rate of amino acid catabolism, so that obligatory losses on a nitrogen-free diet were irrelevant to the estimation of protein requirements. The second conclusion was that, at least for adults, the levels of essential amino acids in dietary proteins could be considerably lower than those present in milk and eggs without their value being reduced. There were fewer new data for young children, but it was judged that, as with adults, the actual requirements (even for milk protein) were approximately 30% greater than the amount calculated from obligatory losses, plus the protein content of growth. However, this change was again largely offset by lower estimates for obligatory losses.[40]

The committee considered that, where there were no reliable values for either the "score" or the biological value of the proteins in particular diets, it would be reasonable to use a "relative protein value" (i.e., relative to a value of 100 for milk protein) of 80% for diets in more affluent countries, where a good proportion of the protein came from animal sources, and of 70% for most of the poorer countries. Where 70% was the relevant value, the safe daily levels of protein intake became 1.82 g/kg for 1-year-olds and 0.82 g/kg for young adults.[41] Since the daily energy requirements of the children and of moderately active young adults were estimated to be 105 and 46 kcal/kg, respectively, the safe protein intakes corresponded in each case to approximately 7 PCals%, respectively. Examples were also given as to how a national balance sheet of protein and energy requirements could be drawn up.

In 1975 another group of scientists was brought together to comment on the 1973 report. Its first criticism was that allowances should have been made for the protein in vegetable foods having a lower digestibility than milk protein; correcting just for differences in "score" based on amino acid composition was inadequate.[42] Also, in practice, habitual energy intakes were for many people less than the assumed energy requirements. For such people the level of dietary protein, expressed as PCals%, would have to be higher than one would calculate from the data given in the report. In addition, there was doubt as to whether the "safe" level of intake of egg

39. Joint FAO/WHO Ad Hoc Expert Committee (1973), p. 49.
40. Ibid., p. 52.
41. Ibid., p. 74.
42. Joint FAO/WHO Informal Gathering of Experts (1975).

protein for adults – that is, 40 g (0.44 × 1.30 × 70) for a 70-kg man – was adequate over a long period. Finally, the group argued that calculation of the adequacy of national food supplies by adding up the requirements of each group within a country could be misleading. Many individuals would choose a dietary pattern with more than the "safe level" of protein. But if the total supply only balanced the total need, it meant that others were inevitably consuming less.

In the same year, Scrimshaw, in a published lecture, discussed the history of committee recommendations on protein requirements. He said that committees were sometimes responsible for serious errors and that "the democratic approach to scientific truth is a contradiction in terms." After reviewing new information on adults' requirements for egg protein, he concluded that the 1973 report (to which he had, of course, been a party) "with its overly confident low estimates of protein requirement has had . . . a variety of undesirable policy, educational, and public-health consequences."[43]

In 1977 there was another small meeting of FAO/WHO consultants. They agreed that calculation of safe intakes of protein should include a correction for digestibility, and in the absence of actual data, a value of 90% could be assumed for developed countries and 85% for others. This would automatically increase the safe intake levels for practical diets by 11% and 18%, respectively. They also attempted to estimate how much protein and energy a stunted child would need to "catch up" to normal size. For a child weighing only 7.5 kg at 12 months of age to reach a normal weight of 11.4 kg at 18 months, the estimated average daily needs (per kilogram body weight) were 1.81 g protein (of milk quality) and 153 kcal (i.e., 4.7 PCals%). If, in addition, the child had an infection, which inhibited growth, for 20% of the time, the estimates increased to 2.41 g protein and 156 kcal (i.e., 6.2 PCals%). Further, if the child, because of individual variability, needed 30% more protein and 15% less energy than average, that child would apparently need a protein concentration of 9.5 PCals%, still in terms of milk protein. Nevertheless, it was thought that, in practice, it was the bulk of the diet and the lack of appetite, rather than protein, which limited a child's possible rate of catch-up growth. Thus a child could be helped by the addition of oil to a diet of cereals and starchy roots, even though this reduced PCals%, because it increased palatability.[44]

The next official FAO/WHO report (published in 1985) was the outcome of a meeting in the fall of 1981. Just before that, a large body of new experimental results had been presented at two conferences sponsored by the UN University World Hunger Program.[45] The report agreed that

43. Scrimshaw (1976), p. 136.
44. Beaton, Calloway and Waterlow (1979).
45. Torun, Young and Rand (1981); Rand, Uauy and Scrimshaw (1984).

more recent experiments indicated that the requirement of young men for egg protein was higher than the estimate of 0.44 g/kg in the 1973 report. In nine studies using single, high-quality proteins, the mean daily requirement for short-term (14-day) nitrogen balance in young men with a habitually higher protein intake was 0.63 g/kg. It was suggested that the difference might be explained by (a) the earlier work having been done generally at energy intakes that promoted weight gain rather than balance, (b) the balance point being extrapolated from points of negative balance in some studies and, finally, (c) the allowance for nitrogen "losses other than in feces and urine" having been set at 5 mg/kg per day rather than the 8 mg used for the newer calculations. In longer (i.e., 1- to 3-month) studies, the mean protein requirement was apparently met by 0.58 g/kg. Therefore, 0.6 g was accepted as the best estimate of the average requirement.[46]

The group then set the safe level of intake of such protein 25% higher (i.e., "+2 standard deviations"), assuming that the between-subject variability was ±12.5%. The safe level of intake, for all but 2.5% of adult men, thus became 0.75 g milk protein per kilogram body weight per day. In the case of adults it was now thought unnecessary to correct for the amino acid score of the diet because in none of a series, even of national dietaries low in animal protein, was the level of any of the essential amino acids limiting when calculated as milligrams amino acid per gram protein and related to the actual amino acid requirements of adults as determined in nitrogen balance studies (Table 10.2). In general, the results of more recent research confirmed the findings of Rose for young adult males, as described in Chapter 11. However, it was agreed that correction was needed where digestibility of the protein was lower than that of egg or milk protein.[47]

For 1- to 2-year-old children, it was considered prudent to add an extra 50% to the allowance for growth, since it was known that growth rate fluctuated from day to day, yet there was little or no storage of dietary proteins in low-growth periods. New research had also indicated that these young children had considerably higher requirements for essential amino acids (per kilogram total protein) than did adults, so that additional protein was recommended where quality was low.[48] The safe allowance corresponded to approximately 10 PCals%. The group confirmed the point made in 1977 that stunted children needing to catch up to a normal size had higher requirements.[49]

46. FAO/WHO/UNU (1985), pp. 79–81.
47. Ibid., pp. 117–20.
48. Torun, Pineda, Viteri and Arroyave (1981).
49. FAO/WHO/UNU (1985), pp. 118–25, 142–8.

Table 10.2. *Comparison of suggested requirement patterns with concentration of four critical amino acids in the protein of low-cost diets*

	Amino acid (mg/g protein)				Scores[a]		
	Lysine	Methionine + cystine	Threonine	Tryptophan	A	B	C
Requirement pattern							
Preschool child	58	25	34	11			
Schoolchild	44	22	28	9			
Adult	16	17	9	5			
Diet type							
Indian diets, vegetarian							
Rice-based	47	31	34	8	73 (Tr)	89 (Tr)	100
Mixed cereals	49	34	33	10	84 (L)	100	100
Tunisian diets, wheat-based							
Rural	33	38	30	11	57 (L)	75 (L)	100
Large town	40	37	32	11	69 (L)	91 (L)	100
Brazilian diets							
Rice/beans/maize	56	30	37	11	97 (L)	100	100
Maize/wheat/cassava/dairy products	56	35	38	11	97 (L)	100	100
Guatemalan diet							
Maize/beans/misc.	39	28	36	10	67 (L)	89 (L)	100
Nigerian diets							
Cassava-based	42	24	30	11	72 (L)	95 (L)	100
Sorghum-based	39	32	33	12	67 (L)	89 (L)	100

[a] The three scores refer to the results of scoring the diet against the assumed requirements of (A) preschool children, (B) schoolchildren and (C) adults. The letters in parentheses indicate which amino acid is apparently limiting when the score is below 100: L, lysine; Tr, tryptophan.
Source: FAO/WHO/UNU (1985).

Is Protein First-Limiting in Third World Diets?

We can now return to the subject, left off at the end of the section titled "Marasmus and Kwashiorkor," as to whether kwashiorkor was to be considered primarily a protein deficiency condition. This had, as we saw, begun to be questioned in the period 1955–7, and increasing emphasis was being placed by some workers on the low energy intakes of children at risk of developing the disease. The greatly reduced estimates of safe levels of protein intake in the 1965 FAO/WHO Report, discussed earlier, also made it seem less likely that kwashiorkor could be considered simple protein deficiency.

In 1959 a study of children's diets in South India indicated, for 1-year-olds, an average daily intake of 1.75 g protein and 76.5 kcal per kilogram body weight from a combination of breast milk and supplements. Compared with the then-current Indian standards, the protein intake was only 50% of requirement and the energy 69%, so that protein appeared to be limiting.[50] By the time a similar study was published in 1969, the Indian Council of Medical Research had modified its standards to bring them in line with FAO/WHO recommendations.[51] As a consequence, although the estimated intakes were similar (i.e., 1.8 g protein and 74 kcal per kilogram body weight daily), it was now concluded that protein intake was adequate and that energy was the factor limiting the growth of these children, who were well below the standards for both height and weight.[52]

P. V. Sukhatme, a statistician working at FAO, calculated, by the same standards, the adequacy of a large number of household diets from data collected in dietary surveys conducted in villages in the state of Madras during the 1950s. In earlier papers he had concluded that these gave evidence of a critical shortage of protein in India.[53] He reanalyzed them against the new standards, which, for a typical household, amounted to a need of 2,800 kcal and 30 g "reference protein" per "nutrition unit," equivalent to 4.3 PCals%. For the analysis he assumed that animal protein in the diets had an NPU of 80% and vegetable protein a value of 50%. The average calculated intakes were 28.5 g "reference protein equivalents" and 2,300 kcal per "unit," that is, 95% and 82% of the protein and energy standards, respectively.

Sukhatme made two further points: first, that the protein standard represented a mean requirement with 2 standard deviations added to allow for individual variability, whereas the energy standard was a mean without any added variability factor; and second, that nearly all of the households whose

50. Rao, Swaminathan, Swarup and Patwardhan (1959).
51. Nutrition Expert Group (1968).
52. Prahlad Rao, Singh, and Swaminathan (1969).
53. Sukhatme (1966), pp. 228–9.

diets were low in protein also had low energy intakes. His final conclusion was that, in India, "the critical factor is the calorie intake and not protein" and that signs of protein deficiency arose because dietary protein had to be used as a source of energy. There was no justification, therefore, for an increased production of protein in all forms regardless of whether these forms were high or low in calories.[54] In a further paper he extended these conclusions, saying that most other national surveys also showed that the average diets had at least 5 PCals% (with protein adjusted to the equivalent level of milk protein), indicating that any protein deficiency would be largely the result of inadequate energy intakes. Under these conditions, supplementing diets with protein or amino acids was costly and inefficient. The only exceptions were those areas where the staples were starchy plantains or cassava.[55]

In late 1970 the FAO/WHO Expert Committee met once more, with Waterlow in the chair. Again, PCM was the subject of a large portion of the report. The committee concluded:

> There has been a tendency to over-emphasize the importance of either protein or calorie deficiency alone, whereas in fact the two almost always occur together. The questions of whether the clinical pictures of kwashiorkor and nutritional marasmus reflect real differences in etiology... have an obvious bearing on diagnosis and on the planning of preventive measures.... Dietary studies fail to provide conclusive evidence. In the "sugar babies" in the West Indies,... the presence of ample amounts of subcutaneous fat suggested that the calorie intake had been relatively high. On the other hand, a study in India of the dietary habits of patients with... nutritional marasmus and kwashiorkor did not show any quantitative or qualitative difference in the previous diets.... In general, the clinical picture of PCM tends to be that of kwashiorkor in regions where breast-feeding is continued into the second year of life, whereas marasmus is the typical result when weaning occurs early.... However, the type of diet available is not always a good indicator of the type of PCM that develops.... We need to obtain more information about the diets of pre-school children, which might serve as a basis for the planning of preventive programmes.[56]

This appears to be a carefully drafted, diplomatic statement, likely to be acceptable to a group with wide differences of opinion.

In 1971 Nicol again wrote about the importance of the bulk factor limiting the nutrient intake of most Third World children and added that, where cereals were the basic staple, the new data supported the old phrase "Look after the calories and the proteins will look after themselves."[57]

Data published the next year from Uganda confirmed that children in their second and third years, with starchy plantains as their staple, and even

54. Sukhatme (1970a).
55. Sukhatme (1970b).
56. Joint FAO/WHO Expert Committee on Nutrition (1971), pp. 51–2.
57. Nicol (1971), p. 85.

when the availability of food was unrestricted, had a mean daily intake equivalent to only 67 kcal/kg body wt, as compared with the standard of 100 kcal. However, to the authors' surprise, most of the Ugandan children showed a normal rate of growth in this period. It was concluded from energy studies that their rates of basal metabolism were similar to those of Western children, but that they were much less physically active, and this was enough to spare at least 20 kcal/kg.[58]

As Whitehead pointed out in a further paper, by comparison with current standards this meant that the apparent energy deficit was greater than the protein deficit, which at 1.5 g/kg and an assumed NPU of 55%, was 71% of the standard.[59] One consequence of the lower energy intake was, of course, that to obtain a standard protein intake the PCals% of the diet had to be that much higher, in inverse proportion – for example, 7.5 PCals% at 67 kcal/kg gives the same protein intake as 5 PCals% at 100 kcal/kg. One could argue that the lower energy intake was, in fact, adequate. However, it left little margin of safety for periods when appetite decreased as a result of infection or other stress. Eating the standard 100 kcal/kg would, for 1- to 2-year-old Ugandan infants, require eating some 1,000 g food, as compared with the 250 g required to provide the same nutrients in a typical Western child's diet. Since a successful catch-up period after an infection would require an even higher intake, it seemed essential to make the Ugandan diet more concentrated, which would be most easily achieved by adding fat.[60] Indeed, the very low fat content of African weaning diets had been thought by one worker to result in a condition of essential fatty acid deficiency.[61]

A further study from India reported the results of giving 300 malnourished preschool children a dietary supplement that provided cereals, fat and sugar, but with a low protein content (4 PCals%), that is, 310 kcal and 3 g protein. The diet of the young children before they received the supplement consisted, on average, of 700 kcal and 18 g protein. Over a 14-month period these children grew significantly faster than 100 control children drawn at random from the same initial pool of subjects. Children receiving the supplement who had the measles showed little or no decline in growth, whereas those in the unsupplemented group who had the measles actually lost weight over the 3 months of the epidemic. The conclusion drawn was that diets based on cereals could be consumed in sufficient quantities if offered in many relatively small meals through the day and that there was no need for the children to receive special protein concentrates; nor did it seem that par-

58. Rutishauser and Whitehead (1972).
59. Whitehead (1973), p. 105.
60. Ibid., pp. 115–17.
61. Naismith (1973).

ticularly high protein levels were needed to provide resistance to infection. The authors commented that people had begun to use the term "protein–calorie malnutrition," but so often still forgot the calorie aspect.[62]

McLaren suggested that "in view of the convincing evidence that had been building up that the main deficiency was energy not protein, the term 'energy–protein malnutrition' should be adopted."[63] To this Scrimshaw and the PAG made a spirited response. They accepted that there was *no* "protein gap" in the sense that the amount of protein available in the world could *theoretically* be distributed so that everyone received enough. However, in the real world the situation was very different, both among families and within families. Further, the diet on which a child developed kwashiorkor was often not its normal diet, but consisted of something like rice (or barley) water given during a long bout of diarrhea. It was accepted that the problem stemmed basically from poverty and ignorance, but it was unacceptable to allow malnutrition to continue until, somehow, in a future generation, poverty was eliminated. The PAG would continue to encourage improved production of protein of all types.[64]

Accusations and Reconsiderations

The battle was now on, and in 1974 McLaren wrote another, more abrasive article for the *Lancet* entitled "The Great Protein Fiasco." He said that the PAG was caught in an identity crisis after having "evolved into a major force in the fight to close the protein gap" and was setting in motion "a long and disastrous train of events." Marasmus was a more widespread disease than kwashiorkor and its incidence was continually increasing as a result of urbanization and the decline in breast feeding. It was because of the biochemical changes, which fascinated laboratory-oriented scientists, that kwashiorkor had received undue attention. Scores of protein-rich food mixtures had been produced; the majority never reached commercial production, and most of the others sooner or later proved financial failures. Several attempts to have the matter discussed at a policy-making level had been thwarted by "the establishment," and criticisms voiced at one seminar had been deleted from the final report, the secretariat "keeping the party line." The true cost in money and time of the numerous projects, countless meetings and innumerable publications was what might have been achieved with the same resources, "whilst children were lost in the unchecked scourge of malnutrition."[65]

62. Gopalan, Swaminathan, Krishna Dumari, Hanumantha Rao and Vijayaraghavan (1973).
63. McLaren (1973).
64. Protein Advisory Group (1973); Scrimshaw (1973).
65. McLaren (1974).

There were some immediate responses. Brock, the South African professor of medicine who had been the co-author with Autret of the 1952 monograph *Kwashiorkor in Africa,* agreed with McLaren's point about the relative neglect of marasmus: "We could have described many undernourished and marasmic African children – they were all around us. But they did not represent what we had come to study and report on." However, he believed that to question the central role of protein deficiency in kwashiorkor was pushing the pendulum too far.[66] Hansen, too, wrote from South Africa. He accepted the point that "the protein-rich food programme had perhaps been over-emphasized." However, he believed that there were important similarities between marasmus and kwashiorkor, and that even a low-energy diet based on cereals was likely to be low in protein, so that McLaren was "perhaps overstating his case and confusing rather than clarifying issues."[67]

For the preceding 20 years, two of the leaders of research groups over the whole period had been Scrimshaw in Guatemala and then at MIT, and Waterlow in Jamaica and then at the University of London. Scrimshaw had already made his position clear on the controversy, but what would be Waterlow's view? It came in 1975 and was also quite clear. In a paper with Payne, published in *Nature,* he withdrew his former position: "The concept of a worldwide protein gap, derived from the diagnosis of kwashiorkor as a protein deficiency state, is no longer tenable: current estimates of children's protein and energy requirements are considered realistic, and by these criteria the problem is mainly one of quantity rather than quality of food."[68] The paper ended: "The evidence we have put forward leads to the conclusion that the protein gap is a myth, and that what really exists, even for vulnerable groups, is a food gap and an energy gap. This point of view is now fairly widely accepted, particularly in this country [Great Britain] where it has become part of aid policy."[69]

The last point is reflected in the report of the Advisory Committee on Protein, which was set up by a British government department to advise on how best to provide overseas aid for the relief of malnutrition. Waterlow was the only member with direct experience in treating malnutrition in the Third World. The summary of the committee's recommendations made no mention of protein. The group advised that "the primary attack on malnutrition should be through the alleviation of poverty . . . in the interests of better nutrition, aid should be directed to projects that will generate income among the poor, even where such projects do not have any marked effect on the national income of the country concerned."[70]

66. Brock (1974).
67. Hansen (1974).
68. Waterlow and Payne (1975), p. 113.
69. Ibid., 117.
70. ODA Advisory Committee on Protein (1975), p. 39.

The report of the World Food Conference organized by the United Nations was also published in 1975. The resolution on "policies and programmes to improve nutrition" extends to some 1,400 words, but "protein" is mentioned only once:

Governments should explore the desirability and the feasibility of meeting nutrient deficiencies, through fortification of staples or other widely-consumed foods, with amino-acids, protein concentrates, vitamins and minerals, and with the assistance of the World Health Organization in co-operation with other organizations concerned, should establish a world-wide control programme aimed at substantially reducing deficiencies of vitamin A, iodine, iron/folate, vitamin D, riboflavin, and thiamine as quickly as possible.[71]

There was no mention of a "world protein gap."

The next report of the FAO/WHO Expert Committee on Nutrition avoided the issue as to whether protein or energy was the main limiting factor in PCM. This was done, it would seem, in part by changing the composition of the committee. Neither Scrimshaw nor Waterlow or Jelliffe was present, and Whitehead and McLaren had never been members. Most of those invited were economists or development planners. According to their report, the evidence indicated that 100 million children were affected by moderate to severe PCM and that "the primary cause of PCM can be overcome only by significant changes in the socio-economic characteristics of the community."[72] They did not state what the primary cause was, but they referred to the report of the preceding meeting, where, in fact, there had been no agreed-upon conclusion. The report then went on, at some length, but in a general way, to recommend that planning be aimed at reducing poverty within countries rather than directly trying to develop "gross national production," which might leave the poor no better off than before, or even worse off. However, there was a comment that "in earlier years... undue emphasis on protein and the so-called 'protein gap' led to undesirable approaches in national and international efforts in the field of food and nutrition."[73]

McLaren had already complained, after the 1971 report of the Joint FAO/WHO Expert Committee, that the series of reports showed prejudices and gross omissions, and that the choice of invitees was political rather than based on scientific merit. "When a group was brought together for only a few days, and with little preparation, a thorough study could not be expected, and it was usually left to the secretariat with one or two of their friends to write the final report."[74] His response to the 1976 report, ex-

71. United Nations (1975), pp. 9–10.
72. Joint FAO/WHO Expert Committee on Nutrition (1976).
73. Ibid., p. 33.
74. McLaren (1972).

pressed in an editorial entitled "Nutrition Planning Daydreams at the United Nations," was even stronger. He described it as being full of undefined, pseudotechnical verbiage concerning the creation of a bureaucratic planning system that would, in some unstated and naively optimistic way, "increase the opportunities for employment." In his opinion, there was a disturbing unwillingness to grapple with the true causes of world hunger and malnutrition in the real world of vested interests and corruption.[75] For whatever reason, the Joint FAO/WHO Expert Committee appeared to go out of existence at this time.

In 1980 a review was published on the role of protein deficiency in kwashiorkor. The authors went into most detail in their consideration of data obtained in Jamaica, but commented that data obtained elsewhere were in agreement with their conclusion. In Jamaica, porridge, primarily cornmeal, was the main weaning food, with about 70 kcal per 100 g. "A typical dietary of 1 bottle of bush tea, 1 bottle of milk, 2 bottles of porridge and 1 meal from the family pot provided 57% of the 1090 kcal required daily between 6 and 11 months of age." If the children were to be given enough porridge to meet their energy standard, the protein intake would be considerably in excess of the RDA. The conclusion was that "there is no basis for the view that kwashiorkor is a protein deficiency disease." The authors stated that the true causes were complex and included infection (especially measles) and possibly a variety of mineral deficiencies, including those of potassium, magnesium, vanadium and zinc, as well as essential fatty acid deficiency.[76] They felt that the bush tea itself might have a damaging effect on a child's liver.[77]

One of the signs of kwashiorkor that seemed particularly likely to be a result of protein deficiency was edema. This had been thought to result from the low levels of serum albumin usually found in infants with kwashiorkor. However, a study from the same group in Jamaica showed that this condition responded to increased energy intake, even when protein was kept down to the daily maintenance level of 0.6 g/kg body wt. The authors said that they did not know why children with kwashiorkor developed edema, but that a low level of protein in the blood did not appear to be the basic cause.[78]

In the 1980s Waterlow, now retired, had some further thoughts on the subject and wrote that he found it hard to believe that edema was *not* due to protein deficiency. Perhaps, after all, the 1973 estimates of protein requirements for children had been set too low, and there was new evidence indicating that the energy requirements had been set unnecessarily high.[79]

75. McLaren (1978), pp. 1296–7.
76. Landman and Jackson (1980).
77. Bras, Berry and György (1957).
78. Golden, Golden and Jackson (1980).
79. Waterlow (1984); Waterlow (1990).

The Complexity of Breast Milk

During this period there was more investigation of the different proteins present in human milk. It was discovered that as much as one-quarter of the total protein consisted of immunoproteins that resisted digestion by gut enzymes and appeared to function as antibacterial agents.[80] The fraction present in the greatest amount consisted of secretory antibodies of the SIgA type, in a form particularly resistant to enzymic attack. It was thought, though there was no rigorous proof, that these proteins played an important role in breast-fed babies generally having fewer episodes of infection than those weaned early and fed in other ways.[81]

Another component found at a higher concentration in human milk than in cow's milk was lactoferrin, a protein with a high affinity for iron. It was hypothesized that this compound acted within the gut to reduce the level of free iron, and thus the proliferation of bacteria that have a high requirement for iron. Other mechanisms helpful to the infant have also been suggested without real proof as to the important function of lactoferrin in vivo.[82]

Finally, it became clear that a significant proportion of infants could develop eczema and other allergic reactions to the proteins in cow's milk. It was an oversimplification, therefore, to think of different protein sources entirely in terms of the amino acids they yielded on hydrolysis. The overall benefits of breast feeding in the Third World and the dangers of mothers thinking it is "modern" or "Western," and therefore that substitutes are preferable, have been argued forcefully by D. B. Jelliffe.[83]

Summing Up

From 1955 on, attention was increasingly focused on the low energy intakes of Third World infants having to subsist on bulky foods. The term "protein–calorie malnutrition" replaced "protein malnutrition," and it was discovered that children recovered from kwashiorkor when given foods of relatively low protein content provided that energy intakes were high. There was controversy as to whether the cause of a malnourished child's developing kwashiorkor rather than marasmus was a lower level of dietary protein or some stress such as infection and/or a failure of hormonal adaptation. In practical terms, prominence was given to ways of increasing the caloric density of Third World infant foods.

80. Räihä (1988).
81. Hanson et al. (1988).
82. Sanchez, Calvo and Brock (1992).
83. Jelliffe and Jelliffe (1978).

It appears that the controversy over kwashiorkor will continue, but with the majority being of the opinion that it is usually the result of the stress of infection in infants receiving overbulky food and that UN programs to meet a world protein crisis by producing protein concentrates were based on a misunderstanding of the situation. Another review of the whole subject was published in 1992.[84]

84. Waterlow (1992).
Note added to proof. It has now been reported that children with kwashiorkor, but not those with marasmus, have greatly increased blood levels of the leukotrienes LTC_H and LTE_4. These compounds stimulate vascular leakage and edema formation. It is suggested that the stresses leading to the development of kwashiorkor result in increased production of these endocrine materials and that they, in turn, contribute to the development of edema [E. Mayatapek, K. Becker, L. Gana, G. F. Hoffman & M. Leichsenrig. (1993). Leukotrienes in the pathophysiology of kwashiorkor. *Lancet* 342:958–60].

11 Adult Needs for Amino Acids: A New Controversy, 1950–1992

It was generally accepted by 1950 that human needs were not really for protein as such but for each of the essential, or "indispensable," amino acids, together with nitrogenous compounds that could be used to synthesize the "nonessential" amino acid units. The latter are those that are not "essential" in the diet because we can synthesize them for ourselves. But determining the adequacy of diets to meet these needs required knowledge of the quantities of individual amino acids supplied by each type of food, as well as reliable information on the requirements of individuals at different stages of life.

In the 1950s methods for analyzing the amino acid composition of foods were greatly improved. Martin and Synge in England had used paper chromatography to separate the amino acids in protein hydrolysates. Then, upon exposure of the paper to the reagent ninhydrin, each amino acid was seen as a colored spot. Moore and Stein in New York developed an automation of the same principle using a chromatographic column.[1] All four won Nobel Prizes for their work in this field.

There was still the question as to how digestible the proteins were, and also whether individual amino acids in some processed foods might undergo chemical reactions that would make them no longer "bioavailable" (i.e., usable by the person or animal consuming them). There was particular concern about possible reactions between the amino acid lysine and the sugars present in a food such as milk powder, and special analytical methods were developed to check for such binding of lysine.[2] Another concern was that excessive levels of one or more amino acids might block the utilization of another. Such an "imbalance" effect could be demonstrated in experiments with rats, but it appeared unlikely to be of significance with the mixtures of foods eaten by humans.[3]

1. Fruton (1972), pp. 165–72.
2. Carpenter (1973).
3. Harper (1974), pp. 155–60.

Amino Acid Requirements

The work of W. C. Rose and his colleagues in developing rat diets that contained free amino acids but no actual protein was described in Chapter 7 together with their first work on the requirements to keep young men in nitrogen balance. Their final conclusions were published as "The Amino Acid Requirements of Adult Man" in 1957.[4] These confirmed that the eight amino acids that they found to be essential were among those needed by the growing rat, but that histidine and arginine were apparently not needed by their subjects.

Similar work was being done in the same period by Ruth Leverton and her colleagues at the University of Nebraska using young women as subjects, and by others.[5] Later work gave some indications of a requirement for histidine.[6] Almost all of the authors referred in their papers to the individual variability of their subjects and to the strange phenomenon in which subjects receiving amino acids in place of dietary protein of the same amino acid composition apparently needed higher energy intakes in order to come into nitrogen balance.

Another unexpected but consistent finding was that, even when the estimates for requirement were based on the highest levels of each essential amino acid that any individual apparently needed to come into nitrogen balance, the total was still quite low relative to estimates of total protein needs. The total daily requirement for all the essential amino acids including histidine came to only about 94 mg/kg body wt. In contrast, the minimum daily protein intake to maintain nitrogen balance was about 600 mg/kg, so that the apparent needs for essential amino acids amounted to only 16% of this value, whereas in the human body, or in whole egg, the essential amino acids make up about 45% of the total protein content. And it had seemed likely that the dietary requirements for replacing broken down tissues in adults would prove to be similar in amino acid composition to that of the tissues being replaced.

In 1957 FAO proposed a hypothetical reference protein for use in scoring the protein of human diets.[7] The essential amino acid requirements amounted in total to 31.4 g per 100 g protein. Histidine was not included, but if it had been the total would probably have been increased by 2 g. This was a compromise between the pattern of infant requirements, which were assumed to be the pattern of amino acids provided by human milk, and the pattern based on nitrogen balance experiments with adults. However, when

4. Rose (1957).
5. Hegsted (1963); Irwin and Hegsted (1971b).
6. Kopple and Swendseid (1975).
7. FAO Committee on Protein Requirements (1957).

the results of a large number of protein-quality trials using nitrogen balance with adult subjects were compared, it was clear that the reference protein pattern did not give reliable predictions. For example, a food with a chemical score (corrected for digestibility) of only 30% showed a performance, based on nitrogen requirements, that was more than 70% that of egg protein.[8]

Successive international committees tried to design a single compromise pattern, but the participants at an international meeting in Rome in 1981 decided to recommend different scoring patterns for different age groups with the adult pattern being based on the results of nitrogen balance experiments summarized by Bodwell.[9] Dietary protein mixtures that had appeared by calculation on the older "chemical score" systems to be seriously deficient in one or more amino acids now appeared to be adequate if consumed by adults at a level providing the safe level of digestible protein (i.e., 750 mg/kg body wt). To take a hypothetical extreme, cooked whole wheat containing 12% protein on a dry-matter basis, if it provided 75% of the total energy needs (i.e., 2,800 kcal) of a 70-kg man and was the only source of protein, would still meet all of the essential amino acid requirements, even though wheat protein had only one-third the lysine context of mixed animal tissue proteins.

"Health Foods" and Amino Acids

In the 1970s amino acids began to appear in health food stores as dietary supplements – ironically, at the same time that scientific work was indicating lower requirements for individual amino acids. The public in Western countries was becoming increasingly nutrition-conscious. After World War II "health food" stores marketing products that catered to the new interest in nutrition proliferated. Previously, "patent" medicines such as "liver pills" with secret formulas and extensive advertising had been widely purchased by people with a vague feeling that they were "below par," but governments had increasingly restricted both the kinds of medicines that could be sold without a physician's prescription and the claims that could be made for their healing or preventive powers. Almost the only products that could now be sold as "purveyors of health" were nutrients, with accompanying claims that the foods in an ordinary diet did not provide them in optimal amounts.

In the 1950s, the common products were mostly vitamin mixtures or combinations of vitamins and minerals. Then health food stores increased their product range to include protein powders (often based on casein),

8. FAO/WHO/UNU (1985), p. 121.
9. Bodwell (1981), p. 350.

which were advertised as a supplement for people wishing to increase their muscle mass or to overcome a feeling of weakness and acquire more energy.

In the late 1960s a "liquid protein" diet was marketed for people who wanted to lose fat but not lean tissue. The protein chosen was unfortunately something like gelatin and was usually partially hydrolyzed to increase its solubility. As we have seen from the earliest work in the nineteenth century, gelatin is not a complete protein; it is, in particular, totally lacking in tryptophan. Several deaths occurred among people trying to live on nothing but "liquid protein," and its production was stopped.[10]

As a consequence, the health food industry became aware of the significance of essential amino acids, and they marketed a range of them, ascribing different virtues to each. The public was encouraged to take one or more as supplements according to the response they desired. An advertisement that appeared in *Prevention* magazine for December 1984 described tyrosine as an "amazing amino acid that resists aging." A flyer found in a California health food shop in 1992 described aspartic acid as "an amino acid for your health.... in studies with human athletes they showed an increase in stamina and endurance... it aids in the disposal of harmful ammonia and thereby helps to avoid damage to the central nervous system." Aspartic acid is one of the amino acids that humans can synthesize for themselves from other nutrients. Glycine was said to assist in prostate health, and lysine was recommended because it was said to lessen the absorption of arginine and thus to stop herpes and other viruses. It was also claimed that one could be more certain of the digestibility of amino acid mixtures than of protein and that these mixtures were available to the tissues in minutes rather than hours.

The material bought most was tryptophan, which was promoted as an aid to people who had a problem sleeping, and apparently it was effective for a proportion of those trying it. However, in 1989 its use was discovered to be associated with the development of a serious nervous disease and even to have been responsible for some deaths. The batches of tryptophan from one supplier were found to be contaminated with a very small quantity of two impurities, each quite similar to tryptophan in structure and not previously suspected of being highly toxic. As a precaution, tryptophan, from any source, was withdrawn from sale.[11]

By 1990 most of the people taking special amino acid mixtures as part of complex supplements were probably body builders. Such supplements were advertised in muscle-building magazines as having "anabolic" (tissue-forming) activity. Some of them, made mostly from intact proteins, were advertised as sources of "sustained-release aminos," a euphemism for

10. Lantigua, Amatruda, Biddle, Forbes and Lockwood (1980).
11. Medsger (1990); Kilbourne (1992).

"slowly digested." By contrast, orthodox scientific opinion held that dietary protein was not the factor normally limiting muscle building, which was under hormonal control.

Nitrogen Balance Reexamined

Returning to orthodox nutrition, the unexpectedness of the low apparent requirements for the essential amino acids led investigators to question the validity of the nitrogen balance procedure as it had ordinarily been carried out. In many studies energy intake had deliberately been kept high to ensure that calories were not limiting, and this had resulted in greater nitrogen retention.[12] Adjustment periods had also usually been inadequate. In addition, there had been either a rather arbitrary allowance for losses from sweat, skin and hair growth or no allowance at all.

Adult protein requirements had not been a high-priority area of research since 1920, because they seemed to be easily met in practice. However, by 1960 NASA, the U.S. space agency, was concerned as to how astronauts remaining in space for long periods could be fed. A mission to Mars and back might take 3 years. Plans were developed for recycling their excreta by redistilling the water present and converting the urea and carbon dioxide back into food, perhaps by photosynthesis, using simple plant organisms. It was essential, in such planning, to have reliable values for the minimum quantities of nutrients that would be needed to constitute a safe, long-term diet. The agency was therefore willing to fund studies that would reexamine the nitrogen balance technique.

Such studies were begun by Doris Calloway, Sheldon Margen and others on the Berkeley campus of the University of California in 1963. A special penthouse unit was set up where six subjects at a time could live under constant supervision for up to 4 months. Only a single food was provided – a premix powder, homogenized with water before each meal and drunk under the eyes of a dietitian. Each subject had a rubber spatula, which he used to scrape out any residues in his mug and then licked clean. The mix included vitamins and minerals, as well as carbohydrate, fat and the protein source being tested. The total energy value was normally adjusted so as to maintain body weight. The standard protein source was powdered egg white.

Physical activity was also controlled, with subjects normally spending 1 hour per day walking on a standard treadmill, but otherwise only sitting or strolling in the unit. The toilets were tied shut, and each subject had his own collection vessels for excreta, which were analyzed. Soiled toilet paper was also analyzed weekly. Skin and sweat losses were measured as follows:

12. Calloway and Spector (1954); Manatt and Garcia (1992).

After showering, each subject was sown into a one-piece cotton suit for the next 3 days. After this, he washed again in a tub with a known volume of water, and the water and the suit were then analyzed for nitrogen. Further analyses were made for ammonia lost in the breath, and for nitrogen in rinsing after toothbrushing and in semen losses. These so-called miscellaneous losses totaled approximately 350 mg nitrogen (equivalent to 2.2 g protein) per day.[13]

Maintaining these procedures was exacting, even sordid, work, but one can see its importance. If, for example, in earlier studies the food provided was estimated to contain 10.0 g nitrogen per day and the collected excreta gave, on analysis, 9.7 g, this result was assumed to constitute an example of positive balance. However, if just 3% of the measured food had really gone as "plate waste" rather than being eaten, and 3% of the excreta had also been lost in being transferred from vessel to vessel, the losses did not cancel out. The true intake of nitrogen was 9.7 g, and the true amount excreted was 10.0 g (i.e., a small negative balance), and if the extra miscellaneous loss was 0.35 g, the true balance was really −0.65 g nitrogen per day, equivalent to 4 g protein. If the efficiency of retention of additional protein at this level of intake had been 50%, it would have required an additional 8 g protein per day to bring the subject into true balance. (There is a problem in presenting results of this kind. Authors refer to losses of milligrams nitrogen per person or per kilogram body weight. However, requirements are thought of in terms of grams protein per head. For consistency I will, from now on, convert nitrogen values to the equivalent weight of protein, that is, "nitrogen content × 6.25," and convert "per kilogram body weight" values to "per head" by assuming 70 kg as the weight of the standard man. Men were used in all the metabolic studies for which results are quoted in this chapter.)

Despite the refinements in carrying out the balance procedure, there was no significant reduction in the apparent variability of subjects. The mean level of endogenous urinary excretion (i.e., after 7 to 10 days on a virtually nitrogen-free diet) was equivalent to 16.6 g protein, with a coefficient of variation (c.v.) of ±18% among individuals.[14] A year later, similar values for altogether 83 young men were reported from a second institute: a mean of 16.2 g protein, with a c.v. of ±15%.[15] The second set of results came from the Massachusetts Institute of Technology (MIT), where Nevin Scrimshaw had moved after his work at INCAP in Guatemala, referred to in Chapters 8 through 10. Of course, the variability of results among individual subjects was not necessarily due to faults in the procedure. There was no

13. Calloway, Odell and Margen (1971).
14. Calloway and Margen (1971).
15. Scrimshaw, Hussein, Murray, Rand and Young (1972).

reason to assume that different individuals would show identical rates of endogenous protein catabolism.

Another concern was that earlier experiments had indicated that there were large, continuing retentions of nitrogen when subjects consumed very large amounts of protein. Again, in a carefully conducted new study, six subjects who received 225 g protein per day had large retentions for the first week, as would be expected, but then apparently continued to retain the equivalent of 10 g/day without any indication of coming back into balance.[16] Could this be a genuine retention, or were some routes of nitrogen loss still not being measured? The question has remained unanswered.

The first practical application of the refined procedure was to determine how far the requirement of subjects for good-quality protein exceeded their endogenous losses. It had been assumed in the 1965 FAO/WHO report on human protein requirements that the requirement for protein of ideal amino acid composition and digestibility could be obtained by summing endogenous losses and, where necessary, adding nitrogen retained for body growth, milk secretion and so on. In other words, there was an assumption of a possible 100% efficiency of conversion from food to tissues.[17] Although the welfare of young children was really the major concern of most investigators, the subjects most readily available for testing the principle were student-aged males.

The results showed that the requirements considerably exceeded the theoretical estimate. The mean daily endogenous losses (urine + fecal + miscellaneous) from a large number of studies were equivalent to 25 g protein, and the mean requirement for balance (using eggs, fish or meat) was 42 g for the standard man.[18] In other words, adults used these first-class proteins with no more than 60% efficiency, whereas they were virtually completely digestible, and on any "scoring" system based on amino acid analyses their scores would have been more than 85%. There was again considerable variability between the calculated requirements of individual subjects, and it was concluded that 52.5 g (or 0.75 g protein per kilogram) from first-class protein would be necessary to meet the needs of 97% of the adult population.[19]

The next question was, what quantity of less well balanced proteins was needed? If a protein had a chemical score of 50 (i.e., 50%) compared with egg protein, did that still mean that twice as much of it would be needed for nitrogen balance? Both the Berkeley and MIT groups carried out relevant studies with protein concentrates made from soybeans.

16. Oddoye and Margen (1979).
17. Joint FAO/WHO Expert Group (1965).
18. FAO/WHO/UNU (1985), pp. 56, 79–80.
19. Ibid., pp. 81–2.

At Berkeley, subjects received varied amounts of either egg white or a soy protein concentrate, with the total nitrogen content of each day's diet made up to the equivalent of 56 g protein with the nonessential amino acids glycine and alanine. It was estimated that the mean minimal amounts of egg and soy proteins required to produce nitrogen balance were 22.5 and 29.5 g, respectively. When the soy protein was supplemented with extra methionine, the minimal requirement was 24.5 g. It appeared, therefore, that the sulfur amino acids were first-limiting in soy for humans, but that even so, soy could be diluted with nonessential amino acids and still maintain nitrogen balance.[20]

In the studies at MIT the protein sources were fed without nonessential nitrogenous supplements. There was again considerable variability among subjects, but when median values were considered, the requirements for nitrogen balance with egg and soy appeared to be 39 and 51 g protein, respectively. Supplementation with methionine did not give consistent improvement, but analyses had indicated that the sulfur amino acid content of the product used in this work was significantly higher than that used at Berkeley.[21] The results from both centers were consistent with the international standard of adult men needing a total of 0.9 g methionine + cystine per day.

Many other nitrogen balance experiments, too many to list individually, have been carried out using particular proteins, with and without amino acid supplements. By and large, they too seem consistent with the international standards for amino acid requirements. However, the results obtained with amino acid mixtures and no protein have been contradictory. In 1955 Rose's group had reported results indicating that they had obtained nitrogen balance (even after adjustment for miscellaneous losses) with two subjects who received the equivalent of 9 g protein from eight essential amino acids and a further 29 g from glycine.[22] They concluded, therefore, that the equivalent of 38 g protein was adequate for a male subject of average weight. Their energy intakes were not stated, but the usual level in Rose's work, as mentioned previously, was about 55 kcal/kg.

Other groups were unable to obtain true nitrogen balance with such a low level of total nitrogen. In one study the six subjects received the same level of total nitrogen and of essential amino acids as in the experiment just described, and 45 to 46 kcal/kg daily.[23] In another study, one treatment provided two-thirds of the quantities of essential amino acids given in the two studies just described (but still 30% more than Rose's estimated re-

20. Zezulka and Calloway (1976).
21. Young, Puig, Queiroz, Scrimshaw and Rand (1984).
22. Rose and Wixom (1955).
23. Kies, Shortridge and Reynolds (1965).

quirement for each) and with additional nonessential amino acids to provide a total equivalent to 44 g protein. The subjects also received approximately 41 kcal/kg. Again, the subjects failed to come into true nitrogen balance.[24] The authors suggested that the failure might have been due to the omission of histidine, which Rose had considered unnecessary. But Rose too had found that his subjects, who received amino acid diets, needed more than 50 kcal/kg to come into nitrogen balance, and it had been repeatedly confirmed since the original work of Voit that nitrogen retention was improved by an increased energy intake.

Amino Acid Turnover Studies

Vernon Young, working at MIT, urged repeatedly that it was unsafe to accept the 1985 international standards for adult requirements for amino acids based solely on the results of nitrogen balance experiments.[25] He had three arguments. First, the procedure itself was suspect because the results indicated that the subjects on a high-protein diet for long periods continued to retain nitrogen. It seemed more likely that this was an artifact and that some nitrogen was actually leaving the body in a form that went unmeasured by the traditional analytical procedures – perhaps as molecular nitrogen in the breath. If this source of error became obvious at high levels of protein intake, it presumably occurred to some degree and without detection at lower intakes. Second, the only attempts to test the full combination of essential amino acids, all included at what Rose had recommended as "safe levels for balance," had in fact not been found to support nitrogen balance in the test subjects. Third, although "balance can be achieved over a range of protein intakes... the rate of protein turnover in cells and organs might well differ.... [Hence] the question that arises is whether a given rate of protein turnover is better than another for health reasons."[26]

Young and his colleagues therefore began a major research program using a new approach to measure indirectly the balance of individual amino acids. Typically the subjects received a powdered diet with energy intakes adjusted so that body weight remained almost constant, as in the nitrogen balance trials described earlier. The daily diets contained no protein but amino acid mixtures contributing nitrogen equivalent to 56 g protein. The amino acid under study was included at different levels, with the other essential amino acids at generous levels and the level of the nonessential aspartic acid adjusted to keep the total nitrogen constant. Usually five or six subjects were used in a study of requirements for a particular amino acid. Each diet was

24. Weller, Calloway and Margen (1971).
25. Young (1986).
26. Ibid., p. 701.

maintained for 7 days, and altogether four or five diets might be tested, each with a different level of the amino acid under study.

After a period of adjustment to a diet, each subject received for several hours in the morning a continuous infusion, into an arm vein, of a saline solution containing the amino acid under study, labeled with the nonradioactive isotope ^{13}C (i.e., carbon with an atomic weight of 13 rather than 12). In some runs the subjects received their food in successive small portions; in others they continued their overnight fast.[27] The rate of expiration of carbon dioxide was measured and portions were analyzed further for their ^{13}C content. At the same time, blood samples were taken and analyzed for their content of the free amino acid under study, and again portions of the amino acid were also isolated and analyzed for their content of ^{13}C. The analyses reached stable plateaus toward the end of 3 hours. The method of calculating balance values from the data obtained in this way is described in Appendix C.

The general conclusion drawn from this work was that the adult requirements for the individual essential amino acids leucine, lysine and threonine appeared to be considerably greater than the standards arrived at as a result of the earlier nitrogen balance studies.[28] Young's conclusion was that the pattern of amino acid requirements of adults (and presumably teenagers also) were not significantly different from the higher standards for 2- to 5-year-olds (shown in Table 10.2) and were close to the actual composition of animal tissues. He believed that this was the only "biologically credible" pattern, in view of the fact that the amino acid pool entering the bloodstream came from the breakdown of body tissues (mostly muscle). Since the proteins of cereals are, in general, considerably lower in lysine than is animal protein, one would expect that people with a marginal intake of protein, most of it coming from cereals, would be deficient in lysine.[29] However, it is clear from experiments with adult rats that their requirement for lysine in relation to other amino acids is very much lower than its relative level in rat tissues, even though rats cannot synthesize it for themselves. These findings have already been described in Chapter 7, and they indicate that the pattern of essential amino acid requirements for maintenance can be very different from the pattern found in animal tissues.

The discrepancies among the results obtained by different procedures are troubling. It is hard to believe that someone could at the same time be in overall nitrogen balance and in negative balance for an essential amino acid, because the amino acid composition of individual proteins is unvarying. The more likely explanation is that there were sufficient errors in one of

27. Meguid et al. (1986).
28. Young and Pellet (1988), pp. 13–18.
29. Ibid., pp. 26–9; Young, Bier and Pellett (1989), pp. 83–9.

the two procedures to distort the conclusions. It has been a pertinent criticism of the nitrogen balance work on amino acid requirements that the energy intakes were excessive, as judged by the usual standards. It would therefore be of great interest to see the results from both types of measurement (i.e., nitrogen balance and isotope turnover) on the same subjects at the same time, so that all the conditions, including energy intake, level of stress, and so on, were identical.

Those who have carried out the newer isotope work have themselves discussed possible sources of error in their own relatively complex studies. However, it would seem that most of these would lead to underestimates of requirements for amino acids so that their correction would only increase the present discrepancies.[30] Another possibility, though it seems remote, is that humans receive significant quantities of indispensable amino acids from the colon, where bacteria can use recycled urea as their nitrogen source. It appears that, in people on a low-protein diet, as much as two-thirds of the urea produced by the liver is secreted into the colon and recycled in one form or another.[31]

Is "Balance" Enough?

One obvious question arising from the MIT studies was whether 7 days were a sufficient period for the subjects to adapt to intakes of an essential amino acid that were considerably lower than those supplied by their previous diets. To test this point, a study was carried out at MIT in which the level of leucine was the variable and each level was ingested for a 3-week period by a group of four or five subjects. One level was 1.0 g/day, corresponding to the 1985 international standard, and another was 2.1 g/day, the level the MIT group had previously found to be necessary to achieve leucine balance. After 1 week of treatment, all the subjects were calculated to be in negative balance, but appeared to have come very close to balance at the end of 3 weeks. However, the authors did not conclude from this that the lower level was satisfactory. Rather, they described the subjects who had received it as being in a state of "accommodation."[32]

This term requires explanation. Waterlow, in particular, has stimulated an interest in the subject of adaptation to changes in the intakes of different nutrients.[33] For example, how is that the rate of urea synthesis adjusts so rapidly to changes in protein intake? He has also made the point that some "adaptations" may be at the expense of full functional ability. A malnour-

30. Meguid et al. (1986), p. 778.
31. Millward, Price, Pacy and Halliday (1991), p. 213; Langran, Moran, Murphy and Jackson (1992).
32. Young, Gucalp, Rand, Matthews and Bier (1987), pp. 14–15.
33. Waterlow (1985).

ished child may survive, but with a reduced rate of growth, and become a "small-sized adult" with less than average muscular mass and physical strength. In this example the "adaptation" has been at the cost of some degree of physical function. Such a result has now been termed "accommodation" to distinguish it from true, or complete, adaptation with no loss of function.[34]

The MIT group's argument that its 3-week results at a relatively low (i.e., 1.0-g) leucine level represented accommodation rather than full adaptation, even though there was "balance," was as follows:

1. On the 1.0-g diet there was a fall in the rate of protein turnover in the second and third weeks of the diet, whereas this did not occur at the 2.1-g level. (This involved further calculations that have not been described here.)
2. When the subjects were returned to a 5.6-g level of intake, there was a period of considerable nitrogen retention for those who had previously received 1.0 g, but much less with those who had received 2.1 g, which indicated that subjects on the 1.0-g treatment had lost a significant quantity of body nitrogen, and there was therefore concern as to their "maintenance of adequate nutriture."[35]

These arguments are analogous to those used by Voit and Atwater 100 years earlier for recommending higher total protein intakes than the minimum levels needed to obtain nitrogen balance. As was discussed in Chapters 5 and 6, they referred to the greater protein "stores" in the bodies of people living on high-protein diets, and they believed that this "labile" protein was a source of nervous energy and drive, as well as of resistance to disease. The authors of the new studies relate the advantage of higher intakes of essential amino acids to increased rates of protein synthesis and, presumably, the potential for more rapid synthesis of immunoproteins, and therefore a more effective response to infection. But to try to measure what is required can lead us in a vicious circle.

The level of protein required for subjects not to lose "labile" protein, and not to show a reduced rate of protein synthesis, is usually the level they were eating before entering the study. This is seen in a study from another center where measurements of amino acid turnover were made just in the first 24-hour period in which the lysine intake was reduced from a high level, in order to determine its requirement by adults.[36] As Hegsted has written of balance studies for calcium, "They represent primarily a study of previous intake."[37]

34. Scrimshaw and Young (1989).
35. Young, Gucalp, Rand, Matthews and Bier (1987), pp. 15–16.
36. Zello, Pencharz and Ball (1993).
37. Hegsted (1952), p. 199.

With the realization that the amino acids released from protein turnover are available for reuse, it is difficult to explain adult needs for amino acids in terms of wear and tear. The nitrogen lost each day in rubbed-off skin, hair and sweat is much less than our minimal total loss, largely as urea. I have suggested elsewhere that, by thinking only of the requirement for protein, we forget our continual need to discard accumulated amino acids.[38] This is achieved by deamination and further metabolism of the carbon fragments. Complete enzyme systems are required to carry out these functions. And since our food supply has throughout our long evolution contained a fair amount of protein, these systems have had to be in continual operation. There would have been no evolutionary advantage in an ability to reduce the activity of the systems further than is actually found. On this view, it would be the detoxification mechanisms for amino acids that determine our minimal requirement rather than any true need of the tissues.

Another worker has argued that "maintenance amino acid needs may be driven in large part by a continuing loss of protein in the gut."[39] This would come from digestive enzymes and other secretions that are not completely digested in the small intestine and are metabolized by bacteria in the large intestine. Amino acids lost in this way come from proteins relatively rich in threonine and the sulfur-containing amino acids.

Unfortunately, we still have no clear data for the relative health, longevity and vigor of peoples whose protein and amino acid intakes differ greatly but whose intakes of trace nutrients are satisfactory and whose environment, life-style and intake of fats, fiber and so on are all generally similar. People living on diets low in protein, or with protein mixtures of doubtful quality, usually have marginal intakes of one or more trace nutrients also. Even the results of animal experiments have been equivocal, and we still lack firm evidence as to whether higher intakes of protein and amino acids confer a long-term advantage.[40]

Fortunately, it appears that even the proposed higher amino acid standards for adults are generally met in practice. The most recent national survey in the United States, using 24-hour recall data, shows a median intake, even for people living below the poverty line, of 35 g protein per 1,000 kcal consumed (i.e., 14 PCals%).[41] This quantity should be ample to meet the proposed higher amino acid standards in view of the relatively high proportion of animal protein in these diets.

There has also been a recent and detailed study of food intake in three villages in different parts of the world where malnutrition problems had

38. Carpenter (1992), pp. 914–15.
39. Reeds (1990), pp. 493–5.
40. Munro (1964), pp. 410–12; Millward (1985), p. 144; Munro (1985), pp. 163–5.
41. U.S. Department of Agriculture (1985).

been expected. The mean daily protein intakes of the adult women subjects in Egypt, Kenya and Mexico, respectively, were 1.1, 1.0 and 1.26 g/kg body wt.[42] Adjusting these values for the calculated digestibility of the proteins and their amino acid score (based on "toddler" standards, as suggested by Young), the quality-corrected values come to 0.95, 0.73 and 0.93 g/kg, respectively; these levels should still meet their requirements.

Upper Limits of Desirable Protein Intake

A concern has arisen in the past few years that high protein intakes may place an undesirable stress on the system. For the first time, the 1989 edition of the U.S. Recommended Dietary Allowances included a recommendation that protein intake not exceed twice the RDA of 0.8 g/kg body wt.[43] The recommended upper limit for the standard man therefore becomes 112 g/day. This highlights the change in views during the twentieth century, since even 115 g was claimed to be insufficient for a working man in official U.S. government reports in 1900.

However, there is no more evidence for 112 g as an upper limit than there was for 115 g being too low. Admittedly, the new recommendation is expressed in tentative terms, but nevertheless once a number gets into the textbooks, it will be assumed that there is hard evidence to support it, as with the high standard of a century earlier. Those who proposed the recommendation felt that higher intakes conferred no benefit, and that there were suggestions from animal experiments that they might accelerate a decline in kidney function.[44] Also, they did not wish to encourage a high level of beef (or other meat) consumption because of the indirect adverse effects of the saturated fat that inevitably accompanies the protein. However, it is a dangerous precedent to assume such possible indirect effects in one direction or the other. In some contexts, as we have seen, meat may also be a vital source of cobalamin (vitamin B_{12}) and zinc.

Summing Up

In the 1950s a series of results were published which indicated that adult requirements for essential amino acids (EAAs) were small, corresponding in total to no more than about 16% of the overall quantity of protein (or amino acids) required to maintain nitrogen balance. This is very different from the 45% of EAAs present in our body proteins. One implication of these results was that protein quality (other than digestibility) would be of

42. Calloway, Murphy and Beaton (1988), p. 22; Calloway et al. (1992), p. 55.
43. National Research Council (1989), pp. 72–3.
44. Committee on Diet and Health (1989), pp. 260–6.

practically no importance in the formulation of adult diets. In practice, virtually any mix of foods that provided the standard quantity of digestible protein would be as good as any other.

However, the correctness of this conclusion depended entirely on the validity of the nitrogen balance procedure. This was simple in principle, but had shown unexplained prolonged retention of nitrogen in subjects continuing to receive diets very high in protein.

A group at MIT has attempted to obtain independent measures of the requirements for individual amino acids by labeling them with stable isotopes and following the balance of the labels. The conclusion drawn from these experiments has been that the requirements are at least double the estimates obtained by the nitrogen balance procedure. Further, feeding the minimal EAA levels required for balance has been found to result in lower turnover rates, and this was thought to be undesirable. The "desirable" levels of EAAs were therefore even higher and roughly equivalent (as percentage of total protein need) to those of toddlers.

This conclusion is still controversial because of the complex assumptions that need to be made when drawing conclusions from studies using isotopes. However, the results of dietary surveys have indicated that, in general, actual adult diets do provide the higher levels of EAAs recommended by the MIT group. A new concern is that people may be consuming an undesirably high intake of protein, which could, with general aging, accelerate the loss of kidney function.

12 In Retrospect

LOOKING BACK AT THE CHANGING IDEAS about protein nutrition described in this book, we see, of course, a great increase in knowledge during each century, from 1700 to 1800, from there to 1900, and then again during the present century. However, on looking at the story in more detail, we see that from decade to decade there was not steady progress, but often "two steps forward and one step back."

It is no surprise for historians of science that in any scientific field ideas which in one generation seemed to be firmly based truths should be overturned in the next. However, it is still a shock for experimental scientists to find that some conclusions which appeared to follow logically from carefully collected data could still turn out to be untrue. It seems to us of prime importance to discover how this could be, so as to be able to reduce, or help our successors to reduce, the danger of drawing wrong conclusions again in the future.

THE COMPLEXITY OF THINGS

One obvious source of error was the assumption of investigators that they had in their hands all the essential information they needed to propose a sensible question and answer it. We saw this after Beccari's extraction of the gluten fraction from wheat flour and the demonstration that it was similar to "animal substance" in its properties. The question was asked: "Which is the nutritive part, the gluten or the starch?" Then because potatoes, rich in starch but lacking gluten, appeared to be a nourishing staple food for peasants in some areas, it was concluded, at least by some, that starch was the essential nutrient. Now we know that it was wrong to conclude that potatoes lacked "animal substance" just because they did not have gluten, which happens to be insoluble and gluey in cold water. Moreover, the question implied as a starting point that there was only one nutrient, an assumption without any experimental basis. In fact, we now realize that nutritional needs are quite complex.

There was a similar error in Takaki's conclusion in the 1880s that the disease beriberi was the result of a deficiency of protein in the diet. As described in Chapter 7, he noted that in the Japanese navy, where the disease was a serious problem, the nitrogen–carbon ratio of the rations issued was well below the German standards for the same period. He therefore hypothesized that this was the cause of the problem, and the Japanese authorities agreed to a trial voyage with foods chosen to provide a higher nitrogen–carbon ratio. The men remained healthy and Takaki understandably believed that he had proved the correctness of his hypothesis. We now know that the foods added to the rations, largely in place of white rice, changed the level of many other nutrients also and that an increase by a few parts per million in the thiamine content of the ration had been the important change. In this instance, the problem had been solved, but the result was actually misleading and confusing to others who had obtained promising results by modifying the procedures used for preparing rice for consumption.

Kwashiorkor was another disease that it was believed at one time could be classified in a simple way as a "protein deficiency" condition, and important allocations of resources were made on that basis. However, further studies have forced workers to see it as another example of a problem whose complexities have not yet been fully unraveled.

There have been a number of studies indicating that, even in communities where there was no obvious malnutrition, the growth of individual children correlated significantly with their intake of animal protein.[1] It is now realized that animal products contribute more than protein, so that "animal protein" could, for example, serve only as an indirect indicator of the level of vitamin B_{12} in the diet. Milk is also a good source of calcium, and meat of iron and zinc. In addition, the minerals from these sources are usually better digested than are those from unleavened grains and vegetables, where they can be bound to fiber and/or phytate.

Falsifying Hypotheses

One idea as to how science progresses, associated particularly with the name of Karl Popper, is that scientists formulate hypotheses, then test and try hard to refute them by carrying out critical experiments. Logically, any number of results may seem to support a generalization but still fail to prove its correctness. Takaki's test of his theory that beriberi was due to protein deficiency is a clear example of this. At the same time, a single result contrary

1. Golden (1988), pp. 153–5; Allen et al. (1992).

to what is predicted by the hypothesis should be sufficient reason to abandon it. But doubts have been raised as to whether this is what happens in practice. Certainly scientists are not usually enthusiastic about trying to refute their own hypotheses, and it is more commonly others who take on that task.

In the protein story, the clearest example of an attempt to refute a hypothesis with a single experiment is that of Fick and Wislicenus in 1865 (as described in Chapter 4). They set out to show that the amount of protein that their bodies broke down and converted to urea during a day's climb would be insufficient to provide the energy required to lift their bodies up the mountain. They expected this to result in a general and immediate abandonment of Liebig's teaching that muscular energy was obtained entirely from the breakdown of a portion of the muscles' own protein tissue.

The results were clear-cut; the authors believed that they had achieved their end and that Liebig would immediately withdraw his hypothesis. In fact, this did not happen. Some scientists were convinced by the work of Fick and Wislicenus, but Liebig and the Munich group "modified" the hypothesis in ways that were thought to preserve its validity: "The broken down protein is not immediately converted to urea, and even when produced there is a delay in its being secreted into the urine: The flaw in the mountain experiment was therefore that urine collection was not continued for a further 24 hours."

When further investigations of the response to physical work showed little evidence of a "second-day" peak in urea secretion, a further modification was suggested: "Muscle protein also breaks down during rest, and this increases the potential energy in the muscle, in a way comparable to the winding up of a spring, that gives it a stored, latent power." Growth of the idea that muscular energy was linked to the oxidation of fats and carbohydrates rather than of protein appears to have been gradual and the cumulative result of many studies, including observations of greatly increased carbon dioxide production during energy expenditure.

As a student of the process of science has written: "Scientists go to great lengths to salvage the views to which they are publicly committed. That way scientific hypotheses get an honest run for their money."[2] In Chapter 6 we saw an example of a scientist being proved right to defend and salvage his views despite an apparent refutation. In 1912 McCay had argued, from a study of different ethnic groups in India, that those receiving the least protein were the least vigorous and that, therefore, Chittenden's recommendation for reducing the protein content of Western diets could also be expected to have a weakening effect in the long term. Chittenden responded by suggesting that the "poor" Indian diets were also lacking in unidentified trace nutrients and that this would prove to be the true basis of their

2. Hull (1988), p. 13.

inadequacy. It seemed a long shot at the time, but subsequent research has lent support to this claim.

In studies of kwashiorkor, one important trial was designed to refute the hypothesis that the disease was caused by some kind of vitamin deficiency. This was the South African work, reported in 1956 (and referred to in Chapter 8), in which sick children were given a mix containing only sugar, minerals and amino acids. The children's positive response was good evidence that kwashiorkor was not due to a deficiency of a vitamin, even an unrecognized one present in milk powders.

After good results were obtained by giving patients mixes with 25% or more of calories coming from milk protein, it became the working assumption that the disease represented a protein deficiency, and there was no deliberate attempt to refute this hypothesis. However, variations in the formulas, designed to raise calorie intakes, had the indirect effect of reducing the proportion of energy coming from protein. Finally, in 1966 (as described in Chapter 10), Chan and Waterlow reported excellent results with a high calorie intake, less than 5% of it coming from protein, and this finding was clearly incompatible with the simple "protein deficiency" hypothesis.

However, I do not believe that any one observation changed the general perception about kwashiorkor. In medicine, individual cases can have special features. The observer therefore develops an "ideal" or "model." As in other fields, "the model represents a simplified structuring of reality . . . [it] gives prominence to some features of our knowledge, and obscures and distorts other features."[3] We hold tenaciously to our models, as providing a satisfying substitute for the overwhelming complexity of the real world. But at some point the accumulation of discordant evidence is sufficient to force us into the unwelcome effort of reorganizing our thoughts and finding a new pattern, or model, for them. With hindsight, it seems that there was a steady accumulation of evidence relating to anorexia and low energy intakes in kwashiorkor from 1955 to 1967. Then suddenly some "last straw" was enough to force a complete revolution of thought, with different pieces of evidence brought to center stage. This is comparable, on a smaller scale, to what Kuhn described as causing a change of paradigms in major scientific revolutions.[4]

The Overextended Generalization

Sometimes a general principle becomes so firmly established in people's minds that they accept it, without question, as a law of nature rather than as a testable hypothesis. The obvious example of this in the history of protein is the principle that only the plant kingdom is capable of building up complex

3. Chorley and Haggart (1967), pp. 22–3.
4. Kuhn (1970).

molecules and that, as a part of the balance of nature, the animal kingdom can only utilize them and/or break them down by oxidation. Both the German and French schools claimed the credit for making this generalization, as we saw at the end of Chapter 2. Undoubtedly this preconception delayed the adoption of iron salts as a simple treatment for the iron-deficiency anemia so common among women in Victorian times. German authorities stated that "just as there was no possibility of humans being able to make protein from potassium nitrate and starch, it was equally improbable that they could use inorganic iron to make hemoglobin."[5]

The belief that animals could not synthesize proteins came partly from the general similarity between proteins found in plants and those in animals. Thus Liebig wrote of "vegetable casein" and "vegetable albumin." However, it was realized in the second half of the nineteenth century that, when proteins were exposed to digestive juices, their properties changed considerably. At first, researchers minimized the importance of this by describing the changes as merely a breakdown of adhesive forces between molecules so that they become more soluble and diffusible across the gut wall. When this no longer seemed an adequate explanation, physiological chemists, as we saw in Chapter 7, spent years trying to categorize the products of gastric digestion as peptones, proteoses and other molecules according to their solubility under different conditions. The unstated reasoning behind this effort was, presumably, that a protein core might be identified that diffused into the bloodstream and formed the basis of our body machinery. When actual crystalline amino acids were recovered from proteins being digested either in vitro or in vivo, they were at first dismissed as artifacts resulting from the abnormal conditions of the experiment. And when the evidence for their existence in the lumen of the small intestine became irresistible, it was argued that this must be nature's way of disposing of surplus protein when the intake exceeded the requirements.

It seems that a general principle can lie so strongly at the back of scientists' minds that it, rather than the observations, is the unconscious starting point of discussion, so that interpretation is not truly open to discussion. To quote another writer, "The way to be objective is not to get along without the concepts and theory, but to get [them]... out into the open, so that nothing is smuggled in surreptitiously and everyone can see precisely what [they]... are."[6]

The "Great Man" Syndrome

Trowell has referred bitterly to Stannus's dismissal, from his Harley Street office in London, of the idea that kwashiorkor could be anything but in-

5. Carpenter (1990), p. 142.
6. Clark (1959), p. 107.

fantile pellagra: "Experts should never be consulted about the status of a new idea in medicine.... they have climbed on the greater steps of their own pinnacle of great eminence; they are hostile to a new idea which may challenge their pre-conceived notions."[7] Such behavior has not been universal, and there are many examples of people reminiscing, in later life, about the help they received from their seniors in their early work. However, in my study of the history of scurvy, it became clear that, time after time, men who had become justly famous for some other discovery were heeded as "authorities" when they gave their opinions that the cause of scurvy was an infection, lack of potassium or of protein and so on, with very little basis for these opinions.[8]

The first obvious example of this phenomenon in relation to protein is Liebig's dictum that protein provided the sole source of the energy required for physical work and that the body could get rid of protein only through the work of muscular contractions. Undoubtedly this statement had great effects on the following generation of teachers. Indeed, it may have been indirectly responsible for thousands of deaths among poor housewives in the southern United States at the beginning of this century who believed that, in addition to giving the available milk to their growing children, they should give most of the meat to their husbands – the "bread winners" engaged in physical work. This left the women with little but cornmeal and white flour, and prey to pellagra, the disease caused by a deficiency of niacin and other B vitamins.[9]

When he published his opinion in 1842, Liebig had already deservedly gained recognition as the leading organic chemist of his day. But as explained in Chapter 3, he had carried out no experimental work in animal physiology, and his elaborate conjectures formed a striking series of hypotheses and nothing more. They carried weight because of Liebig's public reputation, and the energy and ability with which he disseminated them. His critics were outspoken, but their writing appeared only in specialist publications, and the bulk of Liebig's readers did not know of their existence.

Why was Liebig so unwilling to abandon his belief that muscles could obtain energy only from decomposition (or other changes) in their own substance? In part we could say that it was pride and a psychological difficulty in admitting error. Yet he did change his mind on other matters, even if he usually found someone else to blame for having initially misled him. Perhaps the real key lies in his concept of "blood" and his insistence that blood was the only source of nourishment for the tissues. That seems obvious enough, but the picture that a modern biologist has of the blood-

7. Trowell (1982), p. xxii.
8. Carpenter (1986a), pp. 251–2.
9. Carpenter and Lewin (1985), p. 550.

stream is something like a river carrying an enormous assortment of compounds. Some of these, like the immunoglobulins, go around and around, carrying out their function within the bloodstream. In contrast, compounds like glucose and free amino acids are rapidly picked up by some tissues, while other organs introduce more, so that turnover rates bear little relation to the actual concentration of substances in the blood.

All this was unimagined by Liebig when he wrote his *Animal Chemistry*. There was no procedure even for detecting the 0.1% (approximately) of sugar in blood. William Prout himself, who believed that sugar played a special role in the "animal economy," and had specialized in the analysis and identification of sugars, made no mention of it in his table published in 1845 listing the components of blood.[10] It was only in 1847 (after the last edition of *Animal Chemistry*) that Claude Bernard began 10 years of work on the source of the sugar that he found in blood.[11]

Liebig, it seems, thought of blood as a solution (or suspension) of two albuminoids that had essentially the same composition as flesh, so that its conversion, as a whole, seemed only reasonable. Of course, we can think of all sorts of questions that we would like to have asked him, such as "Where did absorbed fats and sugars go, if not into the bloodstream?" But they did not seem to occur to him at the time. Moreover it is understandable that Liebig (and Voit also) found Frankland's analogy with the steam engine difficult to accept. This implied that sugars and fats were burned within the working machinery of the muscles. Yet the evidence at that time indicated that the muscle fibers, which contracted, consisted only of albuminoids and that fats were deposited in separate compartments of the body.

Carl Voit, who succeeded Liebig as the generally accepted "authority" on protein, spent his life doing physiological experiments with both human subjects and dogs. With hindsight, his most important contributions appear to have been his demonstration that increasing total energy intake improved nitrogen balance, and that when an animal in balance was changed to a diet of higher protein content, it first retained nitrogen and then gradually came back into balance but with an increased content of relatively "labile" nitrogen that would be rapidly lost if protein intake was again reduced.

Voit's dictum was that an active man of standard weight should consume 115 g protein per day. But as pointed out at the end of Chapter 5, he never tested whether active men could come into balance and maintain their strength on moderately lower levels (say, 80 to 100 g). The standard was derived entirely from the impression that this was the quantity of protein contained in the food that working men in Voit's area liked to consume when they could afford it, combined with the assumption that when people

10. Prout (1845), pp. 437–9, 467.
11. Holmes (1974), pp. 423–41.

were not limited by poverty they would instinctively select the level of nutrients they needed. Again, we have an example of a standard that the public believed to be authoritative and derived from the scientific work of an expert although, in fact, the standard and the work had little or no connection.

The Contribution of Women

A considerable literature argues that the male scientist's choice of research subjects in biology and the conclusions that he draws tend to be biased toward what are seen as male characteristics of aggression and competition, challenge and defense, to the neglect of cooperation and sharing of resources. In more subtle ways too, male science may have an unconscious bias toward the need for control and mechanization. The obverse of this concept is that women can take a different and complementary approach that may throw an entirely new light on particular topics.

Before 1935 no woman appears in this story. This is to be expected, since scientific research and academic medicine were almost entirely men's worlds. Then Cicely Williams published her first short paper on kwashiorkor. It had never been suggested that this was a new disease in Ghana, so that presumably her attention focused on a problem that was present but not "seen" by her male predecessors in the colonial medical service. As described in Chapter 8, she was publicly snubbed by the current male "authority" on tropical diseases and transferred to Malaysia – presumably to get her quietly out of the way. However, her fate might have been even worse. Her paper was accepted for publication by the editors of the journal to whom she sent it, and men elsewhere in Africa responded in print, stating that they had observed a condition similar to the one she had described.

Twenty years later it had become conventional to categorize kwashiorkor as straightforward protein deficiency and, particularly in the West Indies, to assume that the victims were "sugar babies" receiving, if anything, an excess of calories. At an international meeting on the subject held in Jamaica, a local physician, Katerina Rhodes, claimed that her experience was different: babies that she had seen developing the disease seemed to have little appetite. The regular, international participants at such meetings were reluctant to accept this, but she stood her ground.

Another 2 years later, Dr. Hebe Welbourn, who had been working in child welfare clinics in Uganda, also reported anorexia in her cases, and she reminded readers that Cicely Williams, too, had spoken of poor appetites. I am left with the impression that these women spent more time actually observing the children's behavior in the company of their mothers and nurses – a style contrasting with the caricature (if one likes) of a male doctor hurrying back to the laboratory after gaining a quick clinical impression.

In any case, they clearly made important observations that had not previously received attention.

Committees and Consequences

In the nineteenth century, nutritional standards were put forward by individuals (e.g., Smith, Voit and Atwater), as we have seen. In the 1920s and 1930s, with debates over issues such as whether the poor were suffering from malnutrition serious enough to warrant government intervention, committees began to be established in order to obtain agreement as to what the standards of consumption should be. The period since 1945 has seen the growth of national and international organizations, which are the major sources of funding both for research and for actions designed to help the disadvantaged. It is important that such organizations be seen to be objective and to rely on conclusions agreed to by a committee of experts as a basis for distributing funds. One committee function has been to set out dietary standards of requirements or allowances.

Thanks to rapid air transport, specialists can be assembled from every continent for such committee meetings. We have seen this system at work in the material covered in Chapters 8 through 11. The committees followed in most detail were those organized by the UN system to define the world's nutrition problems and to suggest ways of tackling them. The time period for each meeting has to be settled in advance so that participants can book return flights and plan ahead for the following period. The committee's job, while it meets, typically no more than a week or 10 days, is to produce a report that can be published. This usually means writing with haste in evening sessions and dividing the task among subgroups.

Obviously, senior UN administrators are pleased with a committee that gets the report completed on time and comes up with recommendations that can be carried out within an acceptable budget. Only then does the expenditure on air fares and per diem honoraria seem to be justified. The chair in turn has a heavy responsibility in trying to discipline a group of 8 to 10 senior people, all used to having their opinions regarded as authoritative in their own institutes and free to talk there as long as they like. It is understandable that a chair, in looking over the full list of those qualified, is likely to recommend as committee members people who will probably not haggle over details and will accept majority opinions after a reasonable amount of discussion. The negative side of this is, of course, that the committee is willing to emphasize subjects on which members can agree, so as to give positive answers to sponsors. Conversely, an area of disagreement is deemphasized ("swept under the rug") and referred to in words that, without being actually wrong, still leave a vague impression of overall agreement.

Another characteristic of committee members, particularly when more or less the same group is brought together at fairly frequent intervals, is that they develop a team loyalty and feel individually committed to the opinions expressed in their earlier reports. If these reports are criticized, they naturally want to defend them in an "us-versus-them" situation. Under these circumstances, to defect from the team view is a big decision. We have seen an example of this in Waterlow's dramatic article in 1975, which began with the statement: "It is a most nourishing and stimulating diet to eat one's own words." After chairing one of the first UN committee meetings at which kwashiorkor was classified as a condition of protein deficiency, and then taking the chair at many subsequent meetings with the same theme, he was converted to an opposing view and wrote (with Payne): "The concept of a worldwide protein gap . . . is no longer tenable."[12] One cannot help admiring someone who can say publicly, on reviewing new evidence, "I was wrong."

Scrimshaw has said that "the democratic approach to truth is a contradiction in terms."[13] This is true, and the committee system is not perfect, but as with democracy, we know nothing better. A key point is that minority voices should not be suppressed as a consequence.

The Conduct of Controversy

The question of the reality of the "world protein problem" was brought to a head by Donald McLaren in 1974 with the appearance in the *Lancet* of his article titled "The Great Protein Fiasco."[14] It then became impossible to shrug the matter off as a technical detail relating to a slight difference in emphasis. In that article McLaren directly associated the Protein Advisory Group with "a long and disastrous train of events." This was not the type of language one expects to find in a scientific periodical. McLaren also wrote that, after he had made some criticism of the current concepts at an international meeting, several people said privately that they agreed with him but did not feel free to speak out, fearing that to do so would cause them to lose future research grants. This is a serious point and we will return to it later. But, clearly, McLaren had no such concern. Indeed, a mutual friend has told me that he was using his normal style of speech and correspondence, and that he could not understand anyone taking it amiss. The decision of the editor of the *Lancet* to accept an article written in that style was presumably based on respect for the author's qualifications and experience.

When an eminent Victorian such as Justus Liebig could run his own journal, he had no compunction at being offensive to people he disagreed

12. Waterlow and Payne (1975).
13. Scrimshaw (1976), p. 203.
14. McLaren (1974).

with, even calling them plagiarists or "cocks crowing on a dunghill," as we saw in Chapter 4. The traditional language in British science has continued to be relatively skeptical and direct, with a bite to it. One paper from London referred to the results of the MIT studies, discussed in Chapter 11, as "predictable and irrelevant" to the conclusions being drawn from them. The same group titled another paper "The Winged Bean: Will the Wonder Crop Be Another Flop?" Particular phrases carry different connotations in different cultures. However, it is interesting to find in an editorial review in the *American Journal of Physiology:* "American science has become vitiated by too much politeness.... Conciliatory smoothness is the life blood of diplomacy; it is the death of science. Diplomacy consists of producing agreement (or more usually the illusion of agreement) under a patina of vacuous formulae.... The growing point of science is discrepancy."[15]

The moral perhaps is that we should be as polite as we can be without endangering the clarity of the point we are trying to make. The scientific community should also be careful to include in symposia those with dissident opinions and to ensure that people speaking up will not risk having funding for subsequent investigations disapproved by their peers as a result.

The Responsibilities of Applied Scientists

In recent years, publicity has been given to a few scientists who deliberately falsified data in order to claim the credit for findings that not only were misleading and wasted the time of others, but were just plain dishonest. We all feel it as a stain on our profession that this sort of thing occurs, even rarely. Fortunately, there was no such incident in the present history.

A more common problem is that people become so certain of the truth of a theory or generalization that they select data that fit it and (perhaps unconsciously) reject observations that do not. I am sure that Mulder, for example, truly believed in 1838 that he had discovered that there was a common protein radical present in different albuminoids, as described in Chapter 3. One can also well imagine, in view of the inherent variability in organic analyses at the time, that he regarded those values which agreed best with his preconception as satisfactory, and others that did not as requiring further replication. In the long run, this resulted in his being discredited and in little notice being taken of his later work.

Today there is a more subtle temptation for scientists who have to prepare competitive research proposals in order to obtain funds for their research. This is to look at their task as compiling an argument for support, rather as lawyers on one side of a court case build up the case for their client and omit any contrary evidence on the ground that this is the opposing lawyer's

15. Yates (1983).

responsibility. Papers written for publication have to fit with the preexisting grant application and to act as supporting evidence for the next one, so that even though what is published is "the truth, and nothing but the truth," it may still not be "the whole truth" about a subject, particularly in its promise of practical application. This kind of thing is not often written about, perhaps because people are afraid of giving offense to their peers whose support they need. But one nutritionist has written: "No scientist is simply involved in the single-minded pursuit of truth, he is also engaged in the passionate pursuit of research grants and professional success."[16] This was a courageous statement by John Rivers, who was then at midcareer and could not have known that he had only a few more years to live.

Rivers and his colleagues also criticized a paper in which it was urged that increased funding should be allocated for the development of the winged bean (*Psophocarpus tetrogonolobus*) as a protein-rich food for the tropics. They commented that the dietaries into which the winged bean was to be introduced were not limited in their nutritional value by either protein quality or quantity, and it seemed likely that there would be an overall adverse economic effect on a family unit that adopted the winged bean. They regarded the advocacy of this bean as "merely the latest in a tradition of technological fixes designed to solve the problem of malnutrition . . . [and] the outcome of a continuing process of justifying scientific enthusiasms by the drawing of facile and tenuous links between research which is intellectually exciting to the investigator and problems which are of sufficient public concern to make it politically attractive to devote funds to them."[17]

Some similar criticisms are to be found in a review by one of the leaders of the work done with U.S. federal funding to develop the production of fish protein concentrates: "Much of the motivation for FPC development had little or nothing to do with the ostensible and well-publicized humanitarian goal . . . , it often served as a convenient mask for the varied objectives of the different development groups."[18]

It is a truism in nutrition that most attention has to be paid to the limiting factor, as with the identification of the first-limiting amino acid in a protein. But this principle was not followed in the large body of projects devoted to the development of "single-cell proteins." Early proponents of these materials predicted the quantity of protein that would be produced, by multiplying the nitrogen content of yeast and/or bacteria by the usual 6.25 factor. This neglected the known fact that a considerable proportion of the nitrogen in cells capable of such rapid reproduction is incorporated into nucleic acids rather than into protein. This portion of the nitrogen therefore

16. Rivers (1979).
17. Henry, Donachie and Rivers (1985), p. 337.
18. Pariser, Wallerstein, Corkery and Brown (1978).

represents not an asset but a liability because of the production of uric acid when nucleic acids are ingested by humans. It would have been logical first to have investigated whether these components could be reduced to safe levels by some economical treatment before attempting scaled-up production, but the enthusiasts apparently did not want to raise an issue that might have delayed, or stopped altogether, the chance of their developing new fermenters.

THE PROS AND CONS OF ENTHUSIASM

Poleman, after concluding that FAO's estimates of world food shortages from 1946 to 1974 considerably exaggerated the problem, and that the concept of 500 million people struggling at the brink of starvation was a myth, went on to consider how this had happened. After referring to the possible influence of the special interests of an entrenched bureaucracy and of authors eager to sell books with a dramatic story, he finally took the more generous view that it "reflected nothing more than the persistence of honorable men attempting to dramatize their case through exaggeration."[19]

Nowadays, to describe people as enthusiastic is to praise them for the energy they put into an activity. However, an "enthusiast," when the word was coined in the eighteenth century, was also someone whose ideas were out of perspective or balance. This difference brings us back to the points raised in the preceding section. It was possible for people to be enthusiasts in both the new and the old sense – they were excited about their project and worked hard at it while, at the same time, further consideration might have shown that the program was unlikely to meet the need for which it was designed.

Perhaps the moral is that good intentions are not enough and that we have a duty to be as critical in planning a line of attack on a problem as we will be when actually getting down to work on the selected project. One difficulty here is the specialization of expertise. Food technologists, for example, may reasonably have said that they could not themselves arrive at any responsible opinion as to the problems of the Third World, but that they had read editorial pieces in *Science, Nature* and other respected journals and understood that there was a need for new protein foods, and that that was enough to justify their work on, say, the production of protein isolates from a particular raw material or the attachment of an essential amino acid to a protein source in which this compound was deficient. And in a culture where enthusiasm was honored, and criticism looked down upon as "negative," there was little or no discussion of the value of such work, which

19. Poleman (1977), pp. 386–7.

could continue for a decade before the researchers found that there was no practical use for their product.[20]

Aaron Altschul, who spent many years in basic studies of the proteins in oilseeds and later edited a series of volumes entitled *New Protein Foods*, warned that some attempts at development in the Third World could lead to tragedy, because the new technologies required so many fuel calories for the production of each food calorie. These were commonly not available, and the promise of greater food availability was an illusion.[21]

This brings us back to the complexity of things. We know the kinds of technical problems encountered in our own narrow field, but feel that in other areas, where we are not aware of pitfalls, they probably do not exist, or could easily be overcome. As others have said already, it seemed a simple thing for a protein concentrate that was cheap enough for poultry feeding to be adapted as a human food. And where there was a "need," surely that was virtually the same thing as a "demand." And so on, and so on. Perhaps unconsciously scientists also felt that problems of manufacture, distribution and marketing were relatively trivial once there had been a scientific analysis of what was lacking in a community's food supply.

It seems that, in practice, the problems were great and, moreover, that the original analysis was unbalanced. There are therefore lessons to be learned at all levels. Let us try to apply them, but also to remember in regard to those who seem to have been in the wrong that "the only certain way to avoid making a mistake is never to do anything." We should also have a special respect for the physicians who faced, day after day, the misery of suffering and dying children, and the responsibility of immediately doing the best they could for them.

Some Nutritional Best-Sellers

Just as health food stores sell products, such as amino acids, that orthodox university department nutritionists and physicians are unlikely to recommend, so there is an "alternative" nutritional literature. Some of the best-selling nutrition books are of this kind. The authors convey an impression of absolute certainty as to the correctness of their relatively simple advice and of the almost immediate advantages enjoyed by readers who follow their recommendations. And the public naturally finds these promises appealing.

In many cases these books contain statements that are ludicrous to anyone with even an elementary knowledge of physiology. For example: "It's only undigested food, food 'stuck' in your body that accumulates and becomes

20. Carpenter (1986b), pp. 1367–8.
21. Altschul (1976), pp. 294–5.

fat," and "Once you have eaten a protein in the course of a day, there is no way your body can then digest a carbohydrate," and "Remember that milk is a protein; so that little drop in your coffee will make all the carbohydrates that follow it undigestible or fattening." More than a million copies of the book containing this extraordinary advice were sold, making the author a rich woman.[22] Clearly she knew nothing, or chose to appear to know nothing, about the mechanisms of absorption through the wall of the small intestine that are necessary before nutrients can reach the bloodstream and be circulated to the tissues.

Another best-seller explains that if proteins and carbohydrates are eaten at the same meal, the digestive enzymes nullify one another, since the protein-digesting enzymes require acidic and those digesting carbohydrate need alkaline conditions.[23] Do the authors really not know that a certain amount of digestion occurs in the acidic conditions of the stomach but is completed in the slightly alkaline conditions of the small intestine? The same authors write that the protein in milk makes its calcium unavailable. Again one wonders how they would explain the bone growth of suckling mammals, of whatever species. They also state that the amino acids in egg proteins are unavailable unless the eggs are eaten raw, which is contrary to all actual evidence. Unfortunately, publishers bring such books to the market because they satisfy a demand and yield considerable profit.

Scientists feel more obliged to restrict their claims to those for which there is evidence – and to admit that, while some things are certain, others are less so. They offer some of their recommendations, therefore, on the basis of prudence rather than "proof." Recommendations made with "ifs" and "maybes" obviously have a less forceful impact than a straightforward "Never do this and we guarantee that your skin will recover its youthful appearance in 3 months."

The Bottom Line

Having criticized these popular writings, what recommendations can we offer? Is there a protein problem in the sense that we, and others, are in danger of eating too much or too little? Given the swings in opinions on this subject in the past, it is difficult to be certain that we now know the unalterable truth. However, we do have a much larger body of evidence than our predecessors did, both from direct experiments with human subjects and from field observations in many parts of the world with different dietary traditions. One of the factors that has allowed the human race to

22. Mazel (1981).
23. Diamond and Diamond (1985).

flourish and multiply in so many parts of the planet must be our ability to thrive on many different kinds and combinations of foods.

In the United States 50 g protein per day is officially recommended as an adequate allowance for a healthy woman of average weight (63 kg, or 138 lb.), as is 63 g for a man weighing 79 kg (or 174 lb.). Although there is still controversy over adult requirements for individual amino acids, nearly all the diets that have been studied so far have been found to supply total protein in excess of the RDA and to meet even the highest estimates of amino acid requirements. There seems to be no general need for concern that adults in developed countries are short of protein, or in the other direction that a large proportion are consuming nearly twice their RDA.

Recently there has been more concern whether some intakes are too high, and this is a matter of debate. Certainly many people do consume more than twice the RDA. This is definitely undesirable for people with kidney disease. As discussed earlier, it also seems prudent for others not to exceed this upper limit, even though many do so and seem to be unharmed. Here we enter an area of degrees of risk. Just as we are acquainted with individuals who smoke cigarettes without apparent harm, we know that the proportion of their fellow smokers who have suffered from a number of diseases is very much greater than that of otherwise comparable nonsmokers. The danger from very high protein consumption is much less clear-cut than that from smoking, but it still seems reasonable to recommend caution when there is a possible risk of harm and there is no significant disadvantage to keeping within the limits. Meanwhile, we know that the presently suggested upper limit of "twice the RDA" is a more or less arbitrary one and may well change considerably as knowledge accumulates.

Nutritional advice for relatively affluent adults in recent years has concentrated on reducing fat and cholesterol intake and increasing the consumption of fruit and vegetables. An indirect effect on those following these recommendations has been that their consumption of animal products has declined and that these foods are now served more as flavoring agents in dishes rather than as their major constituents. However, there seems no reason for people to worry that this will leave them short of protein, provided that they eat an otherwise varied diet. Women, who are more at risk than men from osteoporosis in later life, will, in any case, usually want to ensure adequate calcium intake by consuming low-fat milk products that are also good protein sources; they will also need to ensure that they are still receiving an adequate level of iron if they reduce their consumption of meat and do not eat legumes.

Growing children need a slightly higher proportion of protein (to calories) in their diet, but the difference is small because human growth is slower than that of other species and children's energy needs are also high. The first priority is to ensure that they eat enough, and then that their fondness for candy and other sweet things does not prevent them from having a varied, balanced diet. For most children, dairy products again provide a good and palatable source of protein and other nutrients. The small minority who react badly to lactose or cow's milk protein need to have more care paid to their diet. Meat, of course, is a good protein source, and severe fat reduction is not recommended for young children.

"Moderation" and "variety" are still sensible keywords. There is no evidence that we need to go to the extreme of eliminating whole classes of foods or overemphasizing any one class. Our ancestors' "peasant diet" (when things were relatively good) consisted of whole grains, vegetables, seasonable fruit, and a little meat, fish or cheese to impart flavor. The actual choice of grain species and of vegetables depended on their location in the world, but the general pattern reflects long experience and seems difficult to improve upon.

Appendix A: Chemical Structures of Amino Acids

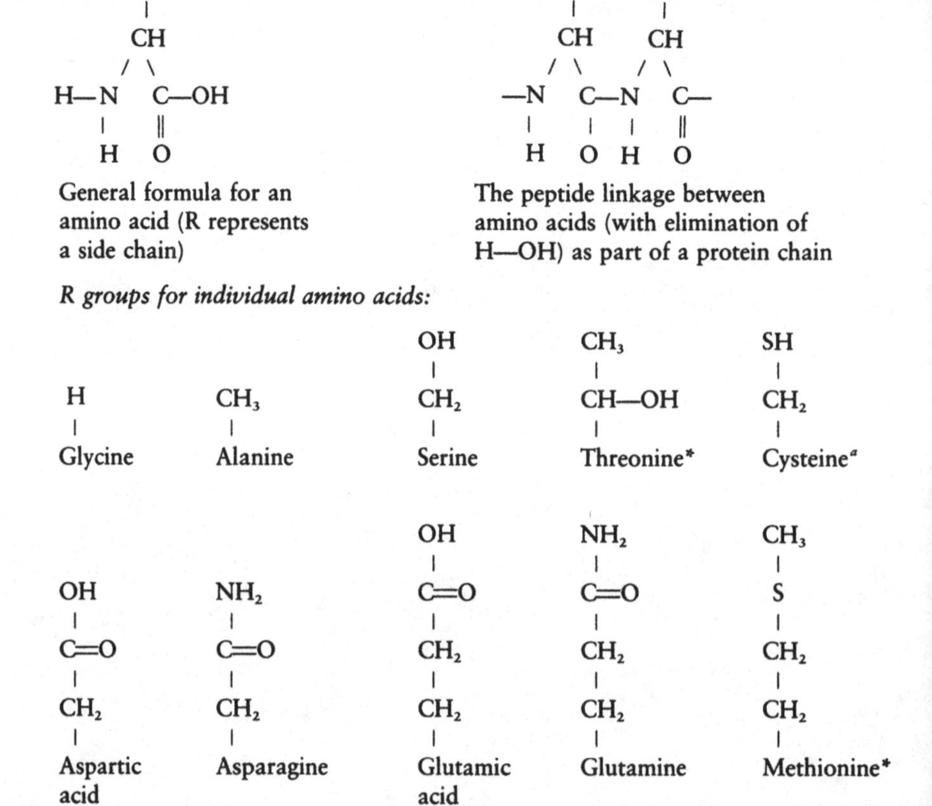

General formula for an amino acid (R represents a side chain)

The peptide linkage between amino acids (with elimination of H—OH) as part of a protein chain

Chemical Structures of Amino Acids

```
                                                                  NH2
                                                                  |
             CH3              CH3                                 C=NH
             |                |                                   |
CH3          CH.CH3           CH2              NH2                NH
|            |                |                |                  |
CH.CH3       CH2              CH.CH3           (CH2)4             (CH2)3
|            |                |                |                  |
Valine*      Leucine*         Isoleucine*      Lysine*            Arginine^b


                                             H   H
                               OH            C— C
                  H            |             /    \
                  C            C           HC      CH
                /   \        /   \          \     /
              HC     CH    HC     CH       HC—CH             H2  H2
C— NH         |      |     |      |         |    \           C —C
||  \         HC     CH    HC     CH        |     NH        /     \
||   CH         \   /        \   /          |    /        H2C      CH
||  //            C            C           C=C              \    / \
C— N              |            |           |   \             N   C—OH
|                 CH2          CH2         CH2  H            |   ||
CH2               |            |           |                 H   O
|
Histidine^b   Phenylalanine* Tyrosine^c    Tryptophan*        Proline^d
```

Note: Asterisks indicate amino acids that are "essential" (or indispensable) because they cannot be synthesized by the body.

[a]"Semiessential," because it can be synthesized from methionine.

[b]Rates of synthesis limited, especially in infants.

[c]"Semiessential," because it can be synthesized from phenylalanine.

[d]Proline does not quite fit the general formula since the tail of the R group is coupled to the NH group of the common structure.

Appendix B: The Measurement of Protein Quality

As DESCRIBED IN CHAPTER 7, it became possible to compare the nutritional values of different protein preparations after protein-free basal mixtures had been developed that supplied all the other nutritional needs of young rats for rapid growth. Osborne and Mendel began their studies by supplying each test protein at different levels, giving their young rats unlimited access to the powdered diets and measuring their weight gains over a 4-week period.

One problem in interpreting the findings was that the diets that supported faster growth had also typically been eaten in greater amounts. Had the rats grown more because they had eaten more, or eaten more because they had grown more? Preliminary trials indicated that rats of a particular size tended to eat similar amounts of their dry diets regardless of the protein source; and in further trials with fixed, limited quantities of food for rats on different diets, growth rates still differed significantly.

Their final procedure was to allow their rats to eat freely (ad libitum) and then to calculate the response as grams weight gained per gram protein eaten, which was later to be called the "protein efficiency ratio" (PER). The value, as one would expect, depended on the percentage of protein in the diet. At very low levels there could be no growth, so that the value had to be zero. Then at very high levels, where more protein was being consumed than was needed for growth, the ratio was again reduced. In general the protein level at which the optimal PER was recorded for a particular protein was higher for poorer proteins – one explanation for this being that more was required before there was a surplus over the maintenance requirement. Some of the relevant data from Osborne and Mendel's paper, published in 1919, is illustrated in Figure B.1. Two protein fractions from cow's milk – lactalbumin and casein – were compared. Lactalbumin was much superior at low levels, but at higher levels the difference almost disappeared.[1]

1. Osborne, Mendel and Ferry (1919).

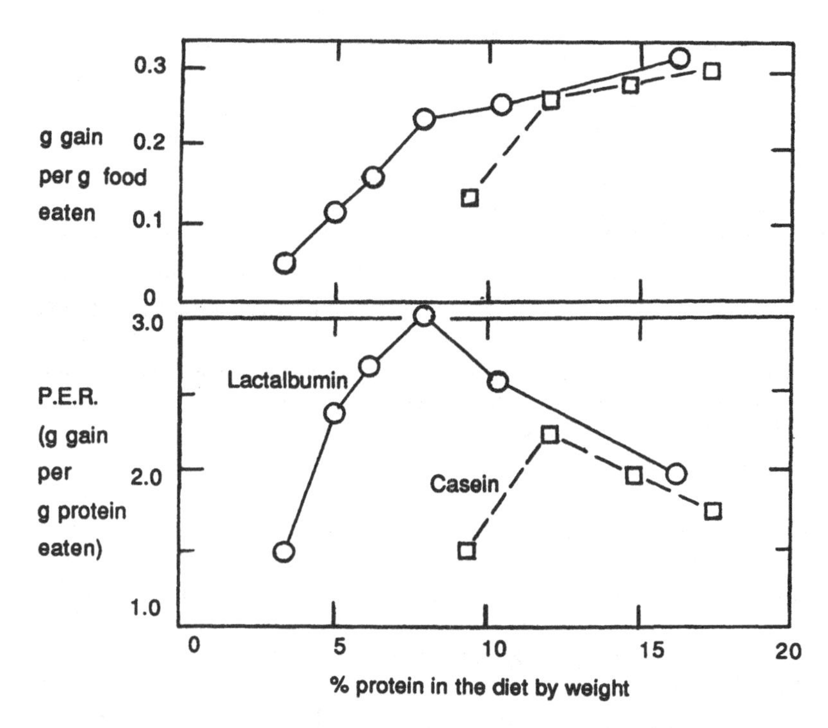

Figure B.1. Mean results from Osborne and Mendel's 1919 paper comparing 4-week protein efficiency ratio (PER) values for casein (□) and lactalbumin (O) when fed at different levels in the diet. (As explained in the text, a 12% protein level here is equivalent, in relation to total calories, to a 9% protein level in the later official procedure. The values also have high standard errors, as many were mean results from only two or three animals.)

At first, therefore, Osborne and Mendel recommended that each protein (or mix) be tested at several levels and that the highest PER value be used as the measure of its quality. However, in practice, workers later used a single protein level (in the range of 9 to 10% of the dry weight of each diet). Some work carried out 25 years later illustrates the general finding that, although this changed absolute values, it did not usually alter the ranking of different proteins (or mixes) for supporting growth in young rats[2] (Table B.1).

Some 50 years after Osborne and Mendel's original work, the PER procedure became an official method of measuring protein quality in the United States.[3] It was specified that, because of the variability from animal to

2. Barnes, Maack, Knights and Burr (1945).
3. Association of Official Agricultural Chemists (1965), pp. 785–6.

Table B.1. *Protein efficiency ratios of four materials*

Protein level	Type of feeding	Test material Whole egg	Soy 1	Soy 2	Wheat gluten
Optimal level for each material	Ad lib	3.8 (100)	2.1 (55)	1.8 (47)	0.8 (21)
10% for all	Ad lib	3.8 (100)	2.0 (53)	1.5 (40)	0.3 (8)
10% for all	Pair-fed[a]	2.6 (100)	1.8 (69)	1.2 (46)	0.4 (15)

Note: Numbers in parentheses indicate percentage of the value for whole-egg protein. Although "whole egg" was used as the standard in this trial, casein was preferred for the official procedure, since it is more stable during storage.
[a]All rats were given the same weight of food.

animal, 10 individually caged rats should be used for each test diet. The level of fat was also reduced to 8%, from the 26% in the early diets. This meant that a diet containing 12% protein by weight had only 9% of the total calories in the form of protein in the early work, and in the later official method the standard 9% by weight also contributed 9% of the energy. This has to be taken into account if one tries to compare old and new results. The specifications also laid down that each trial should include a standard type of casein as one of the test materials. The results with the other test materials could then be standardized by multiplying each value by a number that would bring the result for casein to a value of 2.5. It was thought that this comparison would reduce variability among laboratories that arose from rats in some colonies growing more efficiently than in others.

The method was comparatively simple for a biological test, and useful information was obtained with it, as we shall see later. However, from the beginning the procedure was criticized on a number of grounds: The results depended on an arbitrarily fixed protein level; the tissue gained by the growing rats could vary in its protein content; and no credit was given for a test protein's capacity to provide for the maintenance needs of the animals. Finally, there was no "ideal value" corresponding to a protein that could be utilized with 100% efficiency in replacing worn-out tissue and building new tissue.

The Concept of "Biological Value"

Some of the criticisms of the PER test were overcome in a different type of rat experiment developed by H. H. Mitchell at the University of Illinois's College of Agriculture. Mitchell used a concept first applied by Karl Thomas, working in Rubner's laboratory in Berlin, to results from nitrogen balance experiments with adult human subjects eating different protein foods.

Thomas defined the biological value of the protein in a food as the percentage of the absorbed nitrogen it contributed that was retained – that is, converted to body protein and not reexcreted into the urine – under certain experimental conditions. These were that the protein was not required as a source of energy and that it was not present in great excess of requirement.[4] The method also required estimates of how much nitrogen the experimental subjects would have lost (in feces and urine) if they had received a diet containing no protein, but adequate levels of all other nutrients. Thomas's data are unsatisfactory because his experimental periods were too short to eliminate the effect of the previous treatment. More rigorous data and a discussion of the problems in doing this work with human subjects were presented by another group in 1922.[5]

The calculations involved are best understood by following an example. Imagine that an adult subject on a nitrogen-free diet had a mean fecal nitrogen loss (f) of 1.0 g/day and a urinary nitrogen loss (u) of 4.0 g. Since there was no nitrogen in the diet, the fecal nitrogen must have come from cells rubbed off the wall of the gut and various secretions that were not completely reabsorbed. When the same subject received protein in place of some of the carbohydrate and/or fat in his previous diet and had a nitrogen intake (I) of 7.5 g/day, the fecal loss (F) became 1.5 g nitrogen and the urinary loss (U) 8.0 g nitrogen per day.

Without any dietary protein, the daily balance of incoming and outgoing nitrogen was $0 - 1.0 - 4.0$, that is, -5.0 g. With dietary protein it was $7.5 - 1.5 - 8.0$, that is, -2.0 g. The dietary protein had therefore spared the body reserves by $5.0 - 2.0$, that is, 3.0 g nitrogen per day, and it was concluded that the 7.5 g dietary nitrogen had a "net" value of 3.0 g – a concept comparable to that of a "net profit" calculated after all the accompanying expenses of an operation have been deducted.

The efficiency with which protein has been used can be calculated in terms of the total protein eaten, in which case it is 3.0 as a percentage of 7.5, that is, 40%. (This is now termed the "net protein utilization" or NPU, value of the protein.) Efficiency can alternatively be calculated in relation to the digested nitrogen. Here, since fecal excretion of nitrogen increased by 0.5 g/day when the protein was fed, the absorbed nitrogen is estimated to have been 7.0 g/day, and the 3.0 g net protein is 43% of that quantity. The term "biological value" (BV) is now restricted to calculations made in this way.

For readers who prefer algebraic expressions:

4. Thomas (1909), pp. 238–47.
5. Martin and Robison (1922).

$$\text{NPU} = \frac{I - (U - u) - (F - f)}{I} \times 100$$

$$\text{BV} = \frac{I - (U - u) - (F - f)}{I - (F - f)} \times 100$$

NPU depends both on how well a protein is digested *and* on how well the digested protein is utilized. As one writer has put it, the first factor refers to wastage in digestion and the second to wastage of protein in metabolism. Thus,

$$\text{NPU} = \frac{I - (F - f)}{I} \times \text{BV}$$

Mitchell's Work with Rats

Human studies are slow and expensive. Mitchell, therefore, applied the procedure to rats, with their small food consumption and potential for rapid growth and reproduction.[6] His protocol was to take five individually caged rats (usually young ones weighing about 70 g, but sometimes older ones weighing up to 190 g) and to maintain each of them on the same succession of six diets for 10 days per diet. The first 3 days of each period were allowed for adaptation, and the final 7 days were used to measure nitrogen balance.

The diets for periods 2 through 5 usually contained 8% crude protein (nitrogen content × 6.25) from one of the materials or mixtures being tested, 10% fat (8 parts butterfat, 2 parts cod-liver oil), 4% of Osborne and Mendel's salt mix, 10% sucrose and the remainder starch. In later trials either 4% of cellulose flour or 2% agar was also included, at the expense of starch, to provide roughage.[7] Weighed amounts of food were moistened and put daily into each cage, and the residue 24 hours later recovered, dried and weighed. Usually either 8 or 10 g diet were given each day, depending on the size of the rats at the beginning of each trial. In addition, each rat received a small daily supplement of a yeast extract that was rich in B vitamins, but contributed only 2 to 3 mg nitrogen.

It was the normal practice for the first and last diets to be either nitrogen-free or, in later work, to contain only 4% protein from whole egg, whose nitrogen was considered to be fully digestible, and at this level to replace obligatory nitrogen losses with 100% efficiency, so that urinary nitrogen levels did not increase.[8]

Then the biological value of the test material in each period could be

6. Mitchell (1924a).
7. Mitchell and Carman (1926), pp. 192, 197.
8. Ibid., pp. 192–3.

determined as described earlier. The assumptions were that the fecal nitrogen *not* coming from indigestible food was proportional to the weight of food eaten and that the basal (or endogenous) level of urinary nitrogen was proportional to the body weight of the rat. The appropriate values could thus be obtained by adjusting the values obtained in the first and last periods. Others have expressed doubt as to whether metabolic fecal nitrogen excretion really increased in proportion to food consumption. Since food consumption on the low-protein, or nitrogen-free, diet was always low, not making the adjustment reduces estimated protein digestibility values by about 3 percentage units.[9]

Mitchell and his colleagues carried out many tests of the procedure – varying the order of the treatments, the size of the rats and so on, and increased the preliminary period on each treatment from 3 days to 4. They acknowledged that there were "considerable and consistent individual differences among individual rats in the utilization of dietary nitrogen in metabolism," but believed that mean values based on five rats in each of two trials had a satisfactory degree of precision.[10]

However, when the same materials were fed at levels contributing 5% of the protein in one diet and 10% in another, some very different results were obtained. The biological values for whole milk were 93 and 75%, for corn 72 and 60%, and for potatoes 68 and 67%, respectively.[11] Allowing for maintenance did not therefore always eliminate the effect of "level" even when the protein supplied was not sufficient to provide for optimal growth. The proportions of the proteins used for maintenance and growth, respectively, depended on the level given, and Mitchell explained the higher values obtained for some proteins at the 5% level in terms of the rat not having such critical amino acid requirements for maintenance.

Mitchell published an important review of the complete problem of measuring the nutritive value of proteins in 1924. He referred to 150 papers existing on the subject, but wrote: "Many experimental results defy intelligent interpretation because food intake records are not reported."[12] He concluded:

> The [determined] biological value of a food cannot be considered an absolute value ... it will depend upon the proportion of the absorbed nitrogen that is used for growth ... [and] also upon the level of protein feeding. ... The search for a biological constant seems a vain one. The best that can be hoped for is the determination of values representing fairly the comparative worth of different proteins under certain controlled conditions.[13]

9. Boas-Fixsen (1935), p. 449.
10. Mitchell and Carman (1924), pp. 619–20.
11. Mitchell (1924b).
12. Mitchell (1924c), p. 462.
13. Ibid., p. 443.

Table B.2. *Summary of results obtained with rats by Mitchell's nitrogen balance procedure*

Materials tested	Digestibility[a] (%, mean)	Biological value[a] (%, mean)	Net protein utilization (%, mean)
Eight animal foods	98	80	80
(milk, eggs, meat, offals)	(95–100)	(74–96)	—
Protein fractions from animal foods			
Casein (from milk)	98	69	68
Gelatin (from bones)	96	25	24
Seven legumes and oilseed flours	89	65	58
	(84–97)	(38–80)	—
Seven whole cereal flours	90	66	59
	(85–93)	(54–78)	—

[a]Numbers in parentheses indicate range.
Source: Block and Mitchell (1946).

Some years later, he seemed to reverse this view, writing, "The biological values possess an absolute as well as a relative significance."[14] However, from the context it seems that he merely meant that values for the same material were fairly well reproducible from one trial to another, when the conditions were the same.

After 20 years of work, Mitchell summarized the values obtained in his laboratory and by others using his procedure.[15] Some of the values, collected in groups, are presented in Table B.2. We see that all of the whole animal foods tested are of high protein value, and among these whole eggs and cow's milk, designed by nature as foods for rapidly growing young, are at the top of the list. In contrast, fractions obtained from animal foods can be of lower value. Gelatin, as expected from the findings of early studies, is particularly poor; and casein, obtained by the "curdling" of milk with acid, is significantly poorer than the mixed proteins of whole milk. On average, the vegetable foods tested also have both lower biological values and somewhat lower digestibility than the whole animal foods. However, the mixtures of proteins in whole cereals are superior to isolated proteins such as zein extracted from corn or to the fraction of wheat proteins present in white flour.

14. Mitchell, Burroughs and Beadle (1936), p. 272.
15. Block and Mitchell (1946), p. 263.

Appendix C: Calculations of Amino Acid Balance Using an Isotope Label

THIS IS AN ILLUSTRATIVE SUMMARY of the procedure followed and results obtained in an experiment carried out by Meguid et al. to study human requirements for the amino acid leucine.[1] As stated in Chapter 11, after the subject had been on the test diet for 7 to 21 days, he received an intravenous infusion of the amino acid under study with an isotope incorporated into the molecule.

In this experiment, one subject was given 4 hours of intravenous infusion after 7 days on the test diet. He received 38 mg/hour of leucine in which the number 1 carbon atom (i.e., the carbon atom in the carboxylic acid group) in almost all of the leucine molecules was the nonradioactive isotope ^{13}C. The rate of expiration of CO_2 was measured and portions were analyzed for their ^{13}C content using a mass spectrometer. At the same time, blood samples were taken and their content of free leucine analyzed, and again portions of the leucine were also isolated and their content of ^{13}C analyzed. The analyses reached stable plateaus toward the end of 3 hours. It was assumed that by this time the proportion of ^{13}C in free leucine was the same in tissues where leucine was being oxidized as it was in the blood and that normal and labeled leucine were treated by the body in identical ways.

Simplifying the calculations a little for this run, in which the subject continued to receive food, imagine that the plateau rate of expiration of ^{13}C was 316 μg/hour more than one would expect the total quantity of expired CO_2 to obtain. (The carbon of food and tissues naturally includes a small proportion of ^{13}C, so that it is the "excess" that is calculated.) Other studies had indicated that only 81% of the CO_2 produced from the oxidation of amino acids labeled in this way is recovered in expired breath. Correcting for this, the total ^{13}C coming from [^{13}C]leucine is actually 390 μg, which corresponds to 390/13 = 30 μmol of $^{13}CO_2$. Then if the plasma leucine contained 3% excess ^{13}C, the ratio of total leucine to labeled leucine was

1. Meguid et al. (1986).

100:3. Then, since one would expect the labeled and unlabeled leucine to be oxidized at the same rate, the total quantity of leucine oxidized (with 1 molecule of leucine yielding 1 molecule of CO_2 from its carbon in position 1) is $30 \times 100/3 = 1{,}000$ μmol/hour. Leucine has a molecular weight of 131, so that this corresponds to 131 mg/hour.

In these studies the subjects, on infusion days, usually received their three ordinary meals in the 12 daytime hours, divided up into equal hourly portions for the period of infusion. If their daily diet contained altogether 2,400 mg leucine, each hourly portion contained 200 mg. It could therefore be concluded that at the time of the study the hourly intake of 200 mg exceeded the hourly oxidation of 131 mg, by 69 mg, that is, a situation of positive balance. On the assumption that this hour was typical of the situation during the 12 daytime hours, the total retention in that period would be $69 \times 12 = 828$ mg.

However, to obtain a 24-hour balance, one must have an estimate of nighttime losses when there is no food intake. This was obtained by continuing the subject's overnight fast through the following morning, but otherwise repeating the same infusion experiment. If this yielded a leucine oxidation rate of 115 mg/hour, and if it was assumed to be representative of the 12-hour "night" period, the total oxidation in such a period would be 1,380 mg. Since this exceeds the positive daytime balance of 828 mg, the conclusion would be that the subject was losing 552 mg leucine per 24 hours and that to obtain balance a higher daily intake of leucine would be required than the 2,400 mg provided during this trial.

References

Abderhalden, E. (1905). Abbau und Aufbau der Eiweisskörper im tierischen Organismus. *Z. physiol. Chem.* 44:17–52.

(1909). Weiterer Beitrag zur Frage nach der Verwertung von tief abgebautem Eiweiss im tierischen Organismus. *Z. physiol. Chem.* 61:194–9.

Abderhalden, E., & Rona, P. (1905). Über die Verwertung der Abbauprodukte des Caseins im tierischen Organismus. *Z. physiol. Chem.* 44:198–205.

Abderhalden, E., & Samuely, F. (1905). Beitrag zur Frage nach der Assimilation des Nahrungseiweiss im tierschen Organismus. *Z. physiol. Chem.* 46:193–200.

Achar, S. T. (1950). Nutritional dystrophy among children in Madras. *Br. Med. J.* 1:701–3.

Advisory Committee on the Application of Science and Technology to Development (1968). *International action to avert the impending protein crisis.* New York: United Nations.

Aitken, G. A. (1892). *The life and works of John Arbuthnot.* London: Russell & Russell.

Allen, L. H., et al. (1992). The interactive effects of dietary quality on the growth and attained size of young Mexican children. *Am. J. Clin. Nutr.* 56:353–64.

Altschul, A. M. (1976). Limits of technology. In A. M. Altschul (Ed.), *New protein foods* (Vol. 2, Part B, pp. 280–303). New York: Academic Press.

Altschul, A. M., Lyman, C. M., & Thurber, F. H. (1958). Cottonseed meal. In A. M. Altschul (Ed.), *Processed plant protein foodstuffs* (pp. 469–534). New York: Academic Press.

Anderson, A. (1847). On the recent differences of opinion as to the cause of scurvy. *Monthly J. Med. Sci.* 8:176–81.

Anonymous (1837). Dietetic charlatanry or new ethics of eating. *New York Rev.* 1:336–51.

(1846). Reviews: *Animal Chemistry* by Baron Liebig. *London Med. Gaz., n.s.* 3:1018–24.

(1874). The late Edward Smith, F.R.S. *Br. Med. J.* 2:653–4.

(1886). *Food consumption: Quantities, costs and nutrients of food-materials* (17th Annual Report, pp. 237–326). Boston: Massachusetts Bureau of Statistics of Labor.

(1887a). Kakké or Japanese beri-beri. *Br. Med. J.* 2:189–90.

(1887b). Robert Christison. *Dictionary of national biography* (Vol. 4, pp. 290–1). Oxford: Oxford University Press.

(1906). A discussion on over-nutrition and under-nutrition. *Br. Med. J.* 2:1100–3.
(1911). Discussion on the merits of a relatively low protein diet. *Br. Med. J.* 2:656–67.
(1974). PAG's name changed, scope widened. *PAG Bulletin* 4(3):1–2.
Arbuthnot, J. (1731). *An essay concerning the nature of aliments*. London: Tonson.
Armsby, H. P. (1906). *The principles of animal nutrition* (2d ed). New York: Wiley.
Aronson, N. (1982a). Nutrition as a social problem. *Social Problems* 29:474–87.
(1982b). Social definitions of entitlement: Food needs, 1885–1920. *Media Culture & Society* 4:51–61.
Association of Official Agricultural Chemists (1965). *Official methods of analysis* (10th ed.). Washington, DC: Association of Official Agricultural Chemists.
Atwater, W. O. (1883). *Report of progress of an investigation of the chemical composition and economic value of fish and invertebrates used for food* (Report of the U.S. Commission of Fish and Fisheries for 1880, Part 8, pp. 231–67). Washington, DC: U.S. Government Printing Office.
(1884). Percentages and costs of nutrients in foods (abstract). *Proc. Am. Assoc. Adv. Sci.* (33rd meeting), 648–51.
(1887–8). Pecuniary economy of food. *Century Magazine* 35:437–46.
(1888). Ueber die Ausnützung des Fischfleisches im Darmkanale im Vergleich mit der des Rindfleisches. *Z. Biol.* 24:16–28.
(1891). *American and European dietaries and dietary standards* (4th Annual Report, pp. 106–71). Middletown, CT: Storrs School Agricultural Experiment Station.
Atwater, W. O., & Benedict, F. G. (1902). *Experiments on the metabolism of matter and energy in the human body* (Bulletin No. 109). Washington, DC: U.S. Dept. of Agriculture, Office of Experiment Stations.
Atwater, W. O., & Bryant, A. P. (1900). *Dietary studies of university boat crews* (Bulletin No. 75), Washington, DC: U.S. Dept. of Agriculture, Office of Experiment Stations.
Atwater, W. O., & Langworthy, C. F. (1898). *A digest of metabolism experiments* (Bulletin No. 5), Washington, DC: U.S. Dept. of Agriculture, Office of Experiment Stations.
Atwater, W. O., & Woods, C.D. (1896). *Comments on the dietary studies at Purdue University*. Bulletin No. 32. Washington, DC: U.S. Dept. of Agriculture, Office of Experiment Stations.
(1897). *Dietary studies with reference to the food of the Negro in Alabama*. Bulletin No. 38. Washington, DC: U.S. Dept. of Agriculture, Office of Experiment Stations.
Aulie, R. P. (1970). Boussingault and the nitrogen cycle. *Proc. Am. Phil. Soc.* 114:435–79.
Austin, J. E., Ed. (1979). *Global malnutrition and cereal fortification*. Cambridge, MA: Ballinger.
Autret, M. (1962). Preface to *Encouraging the use of protein-rich foods* (by J. Fridthjof). Rome: Food & Agriculture Organization of the United Nations.
Autret, M., & Behar, M. (1954). *Sindrome policarencial infantil (kwashiorkor) and its prevention in Central America* (FAO Nutritional Studies No. 13). Geneva: United Nations.
Aykroyd, W. R. (1935). *Three philosophers (Lavoisier, Priestley and Cavendish)*. London: Heinemann.

References

Bailey, C. H. (1941). A translation of Beccari's lecture "Concerning Grain." *Cereal Chem. 18*:555–61.

Balland, A. (1902). *La chimie alimentaire dans l'oeuvre de Parmentier*. Paris: Baillière.

Baly, W. (1843). On the prevention of scurvy in prisoners, pauper lunatic asylums, etc. *London Med. Gaz., n.s. 1*:699–703.

Barkas, J. (1975). *The vegetable passion*. New York: Scribner's. (The author may also be catalogued as J. Yager.)

Barker, T. C., Oddy, D. J., & Yudkin, J. (1970). *The dietary surveys of Dr. Edward Smith, 1862–3*. London: Staples Press.

Barnes, R. H., Maack, J. E., Knights, M. J., & Burr, G. O. (1945). Measurement of the growth-promoting quality of dietary protein. *Cereal Chem. 22*:273–95.

Beach, E. F. (1961). Beccari of Bologna: The discoverer of vegetable protein. *J. Hist. Med. 16*:354–73.

Beaton, G. H., Calloway, D. H., & Waterlow, J. (1979). Protein and energy requirements: A joint FAO/WHO memorandum. *Bull. WHO 57*:65–79.

Beaumont, W. (1833). *Experiments and observations on the gastric juice, and the physiology of digestion*. Plattsburgh, NY: Allen (rpt. 1959, New York: Dover).

Beccari, J. (1745). De frumento. *De Bononiensi Scientarium et Artium Instituto atque Academia Commentarii* (Vol. 2, Part 2, pp. 122–7). Bologna: Laelius.

Benedict, F. (1906). The nutritive requirements of the body. *Am. J. Physiol. 16*:409–37.

Berkel, J., & de Waard, F. (1983). Mortality pattern and life expectancy of Seventh-Day Adventists in the Netherlands. *Internat. J. Epidem. 12*:455–9.

Berman, A. (1974). Antoine-Augustin Parmentier. *Dic. Sci. Biogr. 10*:325–6.

Berthollet, C. L. (1785a). Analyse de l'alkali volatil. *Mém. Acad. Sci.*,316–26.

(1785b). Suite des recherches sur la nature des substances animales et sur leurs rapports avec les substances végétales. *Mém. Acad. Sci.*,331–49.

(1786). Sur l'analyse animale comparée à l'analyse végétale. *Observ. Phys. 28*:272–5.

Berzelius, J. J. (1813). Sur la composition des fluides animales. *Ann. Chim. 88*:26–72, 113–41.

(1814). Experiments to determine the definite proportions in which the elements of organic nature are combined. *Ann. Phil.* 4:323–31, 401–9.

(1842). *Jahres-Bericht über die Fortschritte der physischen Wissenschaften*, 22.

Bischoff, T. L. W., & Voit, C. (1860). *Die Gesetze der Ernährung des Fleischfressers*. Leipzig: Winter.

Block, R. J, & Bolling, D. (1945). *The amino acid composition of proteins and foods: Analytical methods and results*. Springfield, IL: Thomas.

Block, R. J., & Mitchell, H. H. (1946). The correlation of the amino acid composition of proteins with their nutritive value. *Nutr. Abstr. Rev. 16*:249–78.

Boas, M. (1956). Acid and alkali in seventeenth century chemistry. *Arch. Internat. Hist. Sci. 9*:13–28.

Boas-Fixsen, M. A. (1935). The biological value of the protein in nutrition. *Nutr. Abstr. Rev.* 4:447–59.

Bodwell, C. E. (1981). Use of amino acid data to predict protein nutritive value for adults. In C. E. Bodwell, J. S. Adkins, & D. T. Hopkins (Eds.), *Protein quality in humans* (pp. 340–69). Westport, CT: Avi.

Boerhaave, H. (1715). *Boerhaave's aphorisms concerning the knowledge and cure of diseases* (J. Delacoste, Trans.). London: Cowse & Innys.

(1719). *A method of studying physick* (Mr. Samber, Trans.), London: Rivington.
Boorde, A. (1567). *A compendyous regyment.* London.
Bouchard, C. (1894). *Lectures on auto-intoxication in disease* (T. Oliver, Trans.) Philadelphia: Davis.
Bouillon-Lagrange, E. F. B. (1805). Analyse de la glu. *Ann. Chim. 56:*24–36.
Boussingault, J. B. (1836). Recherches sur la quantité d'azote contenue dans les fourrages, et sur leurs équivalents. *Ann. Chim. (ser. 2) 63:*225–44.
(1838a). Recherches sur la quantité d'azote contenue dans les fourrages et leurs équivalents: Deuxième mémoire. *Ann. Chim. (Ser. 2) 67:*408–21.
(1838b). Recherches chimiques sur la végétation, entreprises dans le but d'examiner si les plantes prennent de l'azote a l'atmosphère. *Ann. Chim. (Ser. 2) 67:*5–54; *69:*353–67.
(1839a). Analyses comparées des aliments consommés et des produits rendus par une vache laitière: Recherches entreprises dans le but d'examiner si les animaux herbivores empruntent de l'azote à l'atmosphère. *Ann. Chim. (Ser. 2) 71:*113–27.
(1839b). Analyses comparées des aliments consommés et des produits rendus par un cheval soumis à la ration d'entretien: Suite des recherches entreprises dans le but d'examiner si les herbivores prélèvent de l'azote a l'atmosphère. *Ann. Chim. (Ser. 2) 71:*128–36.
(1845). *Rural economy* (G. Law, Trans.) New York: Orange Judd.
Bowie, H. C. (1879). Ueber den Eieweissbedarf eines mitteleren Arbeiters. *Z. Biol. 15:*459–84.
Boyle, R. (1684). *Memoirs for the natural history of humane blood.* London: Smith.
Bras, G., Berry, D. M., & György, P. (1957). Plants as aetiological factor in veno-occlusive disease of the liver. *Lancet 1:*960–2.
Bricker, M., Mitchell, H. H., & Kinsman, G. M. (1945). The protein requirements of adult human subjects in terms of the protein contained in individual foods and food combinations. *J. Nutr. 30:*269–83.
Brock, J. F. (1961). Dietary proteins in relation to man's health. *Proc. Internat. Congr. Nutr.,* 1–5.
(1974). Protein requirement. *Lancet 2:*712–3.
Brock, J. F., & Autret, M. (1952). *Kwashiorkor in Africa* (W.H.O. Monograph Series No. 8). Geneva: United Nations.
Brock, W. H. (1975). William Prout. *Dict. Sci. Biogr. 11:*172–4.
Brock, W. H., & Stark, S. (1990). Liebig, Gregory and the British Association, 1837–1842. *Ambix 37:*134–47.
Broussais, F. J. V. (1826). *A treatise on physiology* (J. Bell, trans.). Philadelphia: Carey & Lea.
Brouwer, E. (1952). Gerrit Jan Mulder (1802–1880). *J. Nutr. 46:*3–11.
Brown, S. C. (1976). Benjamin Thompson (Count Rumford). *Dict. Sci. Biogr. 13:*350–2.
Brown, T. M. (1968). *The mechanical philosophy and the "animal oeconomy."* Ph.D. dissertation, Princeton University.
Browne, C. A. (1942). Justus von Liebig – man and teacher. In F. R. Moulton (Ed.), *Liebig and after Liebig* (pp. 1–9). Washington, DC: American Association for the Advancement of Science.
Bunge, G. (1902). *Textbook of physiological and pathological chemistry* (F. A. Starling, trans.). Philadelphia: Blakiston's.
Bylebyl, J. (1977). Nutrition, quantification and circulation. *Bull. Hist. Med. 35:*470–5.

References

Calloway, D. H., & Margen, S. (1971). Variation in endogenous nitrogen excretion and dietary nitrogen utilization as determinants of human protein requirement. *J. Nutr. 101*:205–16.
Calloway, D. H., Murphy, S. P., & Beaton, G. H. (1988). *Food intake and human function: A cross-project perspective* (Report to US AID). Berkeley: University of California.
Calloway, D. H., et al. (1992). Village nutrition in Egypt, Kenya and Mexico: Looking across the CRSP Projects. Washington, DC: U.S. Agency for International Development.
Calloway, D. H., Odell, A. C. F., & Margen, S. (1971). Sweat and miscellaneous nitrogen losses in human balance studies. *J. Nutr. 101*:775–86.
Calloway, D. H., & Spector, H. (1954). Nitrogen balance as related to caloric and protein intake in active young men. *Am. J. Clin. Nutr. 2*:405–11.
Carman, J. A. (1935). A nutritional disease of childhood. *Trans. Roy. Soc. Trop. Med. 28*:665–6.
Carpenter, K. J. (1973). Damage to lysine in food processing: Its measurement and its significance. *Nutr. Abstr. Rev. 43*:423–51.
(1986a). *The history of scurvy and vitamin C.* Cambridge: Cambridge University Press.
(1986b). The history of enthusiasm for protein. *J. Nutr. 116*:1364–70.
(1990). The history of a controversy over the role of inorganic iron in the treatment of anemia. *J. Nutr. 120*:141–7.
(1991). Edward Smith, 1819–1874. *J. Nutr. 121*:1515–21.
(1992). Protein requirements of adults from an evolutionary perspective. *Am. J. Clin. Nutr. 55*:913–17.
Carpenter, K. J., Ed. (1981). *Pellagra.* Stroudsburg, PA: Hutchinson Ross.
Carpenter, K. J., & Lewin, W. J. (1985). A re-examination of the composition of diets associated with pellagra. *J. Nutr. 115*:543–52.
Carson, G. (1957). *Cornflake crusade.* New York: Rinehart.
Cathcart, E. P. (1912). *The physiology of protein metabolism.* London: Longman, Green.
Chambers, W. H. (1952). Max Rubner, 1854–1932. *J. Nutr. 48*:3–12.
Champagnat, A., Vernet, C., Lainé, B., & Filosa, J. (1963). Biosynthesis of protein–vitamin concentrates from petroleum. *Nature, London 197*:13–4.
Chan, H., & Waterlow, J. (1966). The protein requirement of infants at the age of about 1 year. *Br. J. Nutr. 20*:775–82.
Chapman, C. B. (1967). Edward Smith (?1818–1874), physiologist, human ecologist, reformer. *J. Hist. Med. 22*:1–26.
Charleton [or Charlton], W. (1659). *Natural history of nutrition, life and voluntary motion.* London: Herringham.
Cheyne, G. (1740). *An essay on regimen.* London: Rivington & Leake.
Chibnall, A. C. (1939). *Protein metabolism in the plant.* New Haven, CT: Yale University Press.
Chittenden, R. H. (1895). *On digestive proteolysis.* New Haven, CT: Tuttle, Morehouse & Taylor.
(1904). *Physiological economy in nutrition.* New York: Stokes.
(1907). *The nutrition of man.* New York: Stokes.
(1911). The merits of a relatively low protein diet. *Br. Med. J. 2*:656–62.
Chorley, R. J., & Haggett, P. (1967). *Models in geography.* London: Methuen.
Christison, R. (1847). Account of an epidemic of scurvy which prevailed in the general prison at Perth in 1846. *Monthly J. Med. Sci. 7*:873–91.

Clark, J. T. (1959). The philosophy of science and the history of science. In M. Clagett (Ed.), *Critical problems in the history of science* (pp. 103–40). Madison: University of Wisconsin Press.
Clarke, H. T. (1948). *The use of isotopes in biology and medicine* (pp. 3–15). Madison: University of Wisconsin Press.
Cohnheim, O. (1901). Die Umwandlung des Eiweiss durch die Darmwand. *Z. physiol. Chem.* 33:451–65.
Coleman, W. (1977). *Biology in the nineteenth century.* Cambridge: Cambridge University Press.
Committee on Diet and Health (1989). *Diet and health.* Washington, DC: National Academy Press.
Cosman, M. P. (1983). A feast for Aesculapius. *Ann. Rev. Nutr.* 3:1–33.
Costa, A. B. (1976). Johannes Wislicenus. *Dict. Sci. Biogr.* 14:454–5.
Cowgill, G. R. (1964). Jean Baptiste Boussingault: A biographical sketch. *J. Nutr.* 84:2–9.
Craddock, S. (1983). *Retired except on demand: The life of Dr. Cicely Williams.* Oxford: Green College.
Cravens, W. W., & Sipos, E. (1958). Soybean oil meal. In A. M. Altschul (Ed.), *Processed plant protein foodstuffs* (pp. 353–97). New York: Academic Press.
Crawford, J. (1724). *Cursus medicinae, or a complete theory of physic.* London: Taylor & Osborn.
Crosland, M. (1978). *Gay-Lussac: Scientist and bourgeois.* Cambridge: Cambridge University Press.
Cullen, W. (1789). *A treatise of the Materia Medica.* Philadelphia: Crukshank & Campbell.
Cuthbertson, D. P. (1931). The distribution of nitrogen and sulphur in the urine during conditions of increased catabolism. *Biochem. J.* 25:236–44.
(1942). Post-shock metabolic response. *Lancet,* 1:433–7.
Dally, A. (1968). *Cicely: The story of a doctor.* London: Gollancz.
Dalton, J. (1808). *A new system of chemical philosophy* (rpt. 1964, New York: Philosophical Library).
Darby, W. J. (1973). Cicely D. Williams: Her life and influence. *Nutr. Rev.* 31:331–3.
(1975). Beginnings of PAG. In *The PAG compendium* (Vol. A, v–xxviii). New York: Worldmark Press.
Davies, J. N. P. (1948). The essential pathology of kwashiorkor. *Lancet* 1:317–20.
Davis, A. B. (1973). *Circulation, physiology and medical chemistry in England, 1650–1680.* Lawrence, KA: Coronado Press.
Davis, K. S. (1966). *The cautionary scientists.* New York: Putnam.
Dean, R. F. A. (1949). Ein neuer Vorschlag zur Entwicklung der Kinderernahrung. *Monatsschr. Kinderheirlk.* 97:471–7 (Abstr. 819, *Nutr. Abstr. Rev.* 20 [1950:1]).
(1952). The treatment of kwashiorkor with milk and vegetable proteins. *Br. Med. J.* 2:791–6.
(1953a). Plant proteins in child feeding (Medical Research Council Rep. Special Ser. No. 279). London: H.M. Stationery Office.
(1953b). Treatment and prevention of kwashiorkor. *Bull. World Health Org.* 9:767–83.
(1958). Use of processed plant proteins as human food. In A. A. Altschul (Ed.), *Processed plant protein foodstuffs* (pp. 205–47). New York: Academic Press.

Debus, A. G. (1978). *Man and nature in the Renaissance.* Cambridge: Cambridge University Press.
De Muelenaere, H. J. H. (1969). Development, production, and marketing of high-protein foods. In M. Milner (Ed.), *Protein-enriched cereal foods for world needs* (pp. 266–77). St. Paul, MN: American Association of Cereal Chemists.
Descartes, R. (1649). *A discourse on a method for the well guiding of reason and the discovery of truth in the sciences.* London: Newcombe.
Diamond, H., and Diamond, M. (1985). *Fit for life.* New York: Warner.
Dimino, A. (1969). Incaparina in Colombia. In M. Milner (Ed.), *Protein-enriched cereal foods for world needs* (pp. 341–3). St. Paul, MN: American Association of Cereal Chemists.
Dodart, D. (1731). Mémoire pour servir a l'histoire des plantes. *Mem. Acad. Roy. Sci. depuis 1666 jusqu'à 1699 4:*121–231.
Dombrowski, D. A. (1984). *The philosophy of vegetarianism.* Amherst: University of Massachusetts Press.
Donoso, G., Maccioni, A., & Monckeberg, F. (1967). The use of fish flour as sole source of dietary protein in normal infants. *Proc. 7th Internat. Congr. Nutr., Hamburg 4:*572.
Draper, J. C. (1856). Is muscular motion the cause of the production of urea? *New York J. Med. (Ser. 2) 16:*147–67.
Drummond, J. C., & Wilbraham, A. (1939). *The Englishman's food.* London: Jonathan Cape.
Dubos, R. J. (1950). *Louis Pasteur, free lance of science.* Boston: Little, Brown.
Dumas, J. B., & Cahours, A. (1842). Mémoire sur les matières azotées neutres de l'organisation. *C.R. Acad. Sci. Paris 15:*976–1000.
Edman, M., Forbes, R. M., & Johnson, B. C. (1968). Harold Hansen Mitchell: A biographical sketch. *J. Nutr. 96:*1–9.
Edwards, W. F., & Balzac, _. (1833). Extrait de recherches sur les propriétés de la gélatine. *Ann. Sci. Nat. 26:*318–27.
Eliot, T. (1541). *The castel of health.* New York: Scholars', 1937.
Fantini, B. (1983). Chemical and biological classification of proteins. *Hist. Phil. Life Sci. 5:*3–32.
FAO Committee on Protein Requirements (1957). *Protein requirements* (FAO Nutritional Studies No. 16). Rome: United Nations.
FAO/WHO/UNU (1985). *Energy and protein requirements* (World Health Organization Tech. Rep. Ser. No. 724). Geneva: United Nations.
Fick, A., & Wislicenus, J. (1866). On the origin of muscular power. *Phil. Mag. London (4th ser.) 31:*485–503.
Fidanza, F. (1979). Diets and dietary recommendations in ancient Greece and Rome and the school of Salerno. *Progr. Food Nutr. Sci. 3:*79–99.
Finlay, M. R. (1991). The rehabilitation of an agricultural chemist: Justus von Liebig and the seventh edition. *Ambix 38:*155–67.
(1992). Quackery and cookery: Justus von Liebig's extract of meat and the theory of nutrition in the Victorian age. *Bull. Hist. Med. 66:*404–18.
Fischer, E., & Abderhalden, E. (1903). Über die Verdauung einiger Eiweisskörper durch Pankreasfermente. *Z. physiol. Chem. 39:*81–94.
Fleitman, S. (1986). *Walter Charleton (1620–1707): "Virtuoso."* Frankfurt: Peter Lang.
Flodin, N. W. (1956). The philosophy of amino acid fortification of foods. *Cereal Sci. Today 1:*165–70.

Florkin, M. (1972). *A history of biochemistry.* Vol. 30 of *Comprehensive biochemistry* (M. Florkin & E. H. Stotz, Eds.). Amsterdam: Elsevier.

Food and Nutrition Board. (1958). *Recommended dietary allowances, revised 1958* (Publ. No. 589). Washington, DC: National Research Council.

Forward, C. W. (1898). *Fifty years of food reform: A history of the vegetarian movement in England.* London: Ideal.

Fourcroy, A. F. de (1789a). Recherches pour servir à l'histoire du gaz azote ou de la mofette, comme principes des matières animales. *Ann. Chim. 1*:40–6.

(1789b). Mémoire sur l'existence de la matière albumineuse dans les végétaux. *Ann. Chim. 3*:252–62.

(1792). Axiômes chimiques. In *Encyclopédie méthodique* (Vol. 2, pp. 455–89). Paris: Panckoucke.

Fowler, O. S. (1871). *Physiology, animal and mental* (5th ed.). New York: Fowler & Wells.

Frankland, E. (1866a). On the origin of muscular power. *Phil. Mag. London (4th ser.) 32*:182–99.

(1866b). On the source of muscular power. *Roy. Inst. Proc. 4*:661–85.

Fruton, J. S. (1972). *Molecules and life.* New York: Wiley-Interscience.

(1979). Early theories of protein structure. *Ann. N.Y. Acad. Sci. 325*:1–15.

Funk, C. (1912). The etiology of deficiency diseases. *J. State Med. 20*:341–68.

(1913). Studies on beri-beri: 7. Chemistry of the vitamine-fraction from yeast and rice-polishings. *J. Physiol. 46*:173–9.

Garrison, F. H. (1929). *An introduction to the history of medicine* (4th ed.). Philadelphia: Saunders.

Garrow, J. S., Smith, R., & Ward E. E. (1968). *Electrolyte metabolism in severe infantile malnutrition.* Oxford: Pergamon.

Gelbart, N. R. (1971). The intellectual development of Walter Charleton. *Ambix 18*:149–68.

Gelfand, M. (1946). Kwashiorkor. *Clin. Proc. J. Cape Town Post-Grad. Med. Assoc. 5*:135–53. (Abstr. 4744, *Nutr. Abstr. Rev. 16* [1947].)

Gillman, J., & Gillman T. (1951). *Perspectives in human malnutrition.* New York: Grune & Stratton.

(1945). Hepatic damage in infantile pellagra, and its response to vitamin, liver and dried stomach therapy as determined by repeated liver biopsies. *J. Am. Med. Assoc. 129*:12–9.

Glas, E. (1975). The protein theory of G. J. Mulder (1802–1880). *Janus 62*:289–308.

Golden, M. H. N. (1988). The role of individual nutrient deficiencies in growth retardation of children as exemplified by zinc and protein. In J. C. Waterlow (Ed.), *Linear growth retardation in less developed countries* (pp. 143–63). New York: Raven Press.

Golden, M. H. N., Golden, B. E., & Jackson, A. A. (1980). Albumin and nutritional oedema. *Lancet 1*:114–6.

Goodman, D. C. (1971). The application of chemical criteria to biological classification in the eighteenth century. *Med. Hist. 15*:23–44.

Gopalan, C. (1967). Major nutritional problems of India and Southeast Asia. *Proc. 7th Internat. Congr. Nutr., Hamburg 3*:320–6.

(1968). Kwashiorkor and marasmus: Evolution and distinguishing features. In R. A. McCance & E. M. Widdowson (Eds.), *Calorie deficiencies and protein deficiencies* (pp. 49–58). London: Churchill.

Gopalan, C., & Patwardhan, V. N. (1951). Some observations on the "nutritional oedema syndrome." *Ind. J. Med. Sci. 5*:312–7.

Gopalan, C., Swaminathan, M. C., Krishna Dumari, V. K., Hanumantha Rao, D., & Vijayaraghavan, K. (1973). Effect of calorie supplementation on growth of undernourished children. *Am. J. Clin. Nutr. 26*:563–6.

Gordon, H. H., & Ganzon, A. F. (1959). On the protein allowances for young infants. *J. Pediatr. Sci. 54*:503–28.

Graham, G., Cordano, A., & Baertl, J. M. (1963). Studies in infantile malnutrition: 2. Effect of protein and calorie intake on weight gain. *J. Nutr. 81*:249–54.

Graham, G. G., & Morales, E. (1963). Studies in infantile malnutrition: 1. Nature of the problem in Peru. *J. Nutr. 79*:479–87.

Graham, S. (1839). *Lectures on the science of human life* (2 vols.). Boston: Marsh, Capen, Lyon & Webb.

Grande encyclopédie (1899). Vol. 25, pp. 1177–8.

Greenstein, J. P., & Winitz, M. (1961). *Chemistry of the amino acids* (3 vols.). New York: Wiley.

Grmek, M. D. (1974). François Magendie. *Dict. Sci. Biogr. 9*:6–11.

Guerlac, H. (1961). *Lavoisier – the crucial year: The background and origin of his first experiments on combustion in 1772*. Ithaca, NY: Cornell University Press.

Guggenheim, K. Y. (1981). *Nutrition and nutritional diseases: The evolution of concepts*. Lexington, MA: Heath.

(1991). Rudolf Schoenheimer and the concept of the dynamic state of body constituents. *J. Nutr. 121*:1701–4

Guibourt, M., & Depaul, M. (1867). Sur un lait artificiel. *Bull. Acad. Méd. Paris. 32*:803–8.

Hall, T. S. (1969). *Ideas of life and matter* (2 vols.) Chicago: University of Chicago Press.

Hallé, J. N. (1791). Essai de théorie: Sur l'animalisation et l'assimilation des alimens. *Ann. Chim. 11*:158–74.

Haller, A. (1754). *Dr. Albrecht Haller's physiology* (2 vols.). London: Innys & Richardson.

Hammond, E. G. (1987). Sylvester Graham: America's first food reformer. *Iowa St. J. Res. 62*:43–61.

Hammond, W. A. (1855). The relations existing between urea and uric acid. *Am. J. Med. Sci. 29*:119–23.

Hansen, J. D. L. (1968). Features and treatment of kwashiorkor at the Cape. In R. A. McCance & E. M. Widdowson (Eds.), *Calorie deficiencies and protein deficiencies* (pp. 33–47). London: Churchill.

(1974). Protein requirement. *Lancet 2*:713–14.

Hansen, J. D. L., Howe, E. E., & Brock, J. F. (1956). Amino-acids and kwashiorkor. *Lancet 2*:911–13.

Hanson, L. Å., et al. (1988). Antiviral and antibacterial factors in human milk. In L. Å. Hansen (Ed.), *Biology of human milk* (pp. xx–x). New York: Raven Press.

Harper, A. E. (1974). Effects of disproportionate amounts of amino acids. In A. E. Harper & D. M. Hegsted (Ed.), *Improvement of protein nutrition* (pp. 138–66). Washington, DC: National Academy of Sciences.

Harris, H. F. (1919). *Pellagra*. New York: Macmillan.

Hegsted, D. M. (1952). A study of the minimum calcium requirements of adult men. *J. Nutr. 46*:181–99.

(1963). Variation in requirements of nutrients: Amino acids. *Fed. Proc. 22*:1424–30.

Hegsted, D. M., Tsongas, A. G., Abbott, D. B., & Stare, F. J. (1946). Protein requirements of adults. *J. Lab. Clin. Med. 31*:261–84.

Heilbron, J. (1991). The contribution of Bologna to galvanism. *Histor. Stud. Phys. Biol. Sci. 22*:57–85.

Helmholtz, H. (1861). On the application of the law of force to organic nature. *Roy. Inst. Proc. 3*:347–57.

Helmont, J. B., van. (1662). *Oriatrike or physick refined* (J. Chandler, Trans.). London: Loyd.

Henriques, V., & Hansen, C. (1904). Ueber Eiweisssynthese im Thierkörper. *Z. Physiol. Chem. 43*:417–46.

Henry, C. J. K., Donachie, P. A., & Rivers, J. P. W. (1985). The winged bean: Will the wonder crop be another flop? *Ecol. Food. Nutr. 16*:331–8.

Hindhede, M. (1913). *Protein and nutrition*. London: Ewart, Seymour.

Hirschfeld, F. (1887). Untersuchungen über den Eiweissbedarf des Menschen. *Pflügers Arch. 4*:533–66.

Holemans, K., Lambrecht, A., & Martin, H. (1956). Résultats d'une campagne d'alimentation supplémentaire par protéines végétales (arachides). *Mém. Acad. Sci. colon. Classe Sci. nat. méd. (n.s.)*4 (6).

Holmes, F. L. (1963). Elementary analysis and the origins of physiological chemistry. *Isis 54*:50–81.

(1964). Introduction to *Justus Liebig's "Animal Chemistry"* (pp. vii–cxvi). New York: Johnson Reprint.

(1971). Analysis by fire and solvent extractions: The metamorphosis of a tradition. *Isis* 62:129–48.

(1973). Justus von Liebig. In *Dict. Sci. Biogr. 8*:329–50.

(1974). *Claude Bernard and animal chemistry*. Cambridge, MA: Harvard University Press.

(1975). The transformation of the science of nutrition. *Isis 8*:135–44.

(1979). Early theories of protein metabolism. *Ann. N.Y. Acad. Sci. 325*:171–85.

(1985). *Lavoisier and the chemistry of life*. Madison: University of Wisconsin Press.

(1987). The intake–output method of quantification in physiology. *Hist. Stud. Phys. Sci. 17*:239–70.

Homberg, M. (1702). Essais de chymie. *Mém. Acad. Sci.*, 33–52.

Hopkins, F. G., & Cole, S. W. (1902). A contribution to the chemistry of proteids: 1. A preliminary study of a hitherto undescribed product of tryptic digestion. *J. Physiol.* 27:418–28.

Hudson, R. P. (1989). Theory and therapy: Ptosis, stasis and autointoxication. *Bull. Hist. Med.* 63:392–413.

Hull, D. L. (1988). *Science as a process*. Chicago: University of Chicago Press.

Icaza, S. J. (1976). Case histories in mass communications: The Incaparina project. *PAG Bull.* 6(1):35–7.

Idler, D. R. (1969). The Halifax isopropanol process for the manufacture of FPC. In *Production of fish protein concentrate* (pp. 107–15). New York: United Nations.

Ihde, A. J. (1964). *The development of modern chemistry*. New York: Harper & Row.

INCAP. (1962). *Incaparina*. Guatemala City: Institute of Nutrition of Central America and Panama.

Irwin, M. I., & Hegsted, D. M. (1971a). A conspectus of research on protein requirements of man. *J. Nutr. 101*:385–430.

(1971b). A conspectus of research on the amino acid requirements of man. *J. Nutr. 101*:539–66.

Jackson, R. W., & Block, R. J. (1932). The availability of methionine in supplementing a diet deficient in cystine. *J. Biol. Chem. 98*:465–77.

Jägerroos, B. H. (1902). Ueber die Folgen einer ausreichenden, aber eiweissarmen Nahrung. *Skand. Arch. Physiol. 13*:375–418.

Jaya Rao, K. S. (1974). Evolution of kwashiorkor and marasmus. *Lancet 1*: 709–11.

Jaya Rao, K. S., Srikantia, S. G., & Gopalan, C. (1968). Plasma cortisol levels in protein–calorie malnutrition. *Arch. Dis. Child. 43*:365–7.

Jelliffe, D. B. (1959). Protein–calorie malnutrition in tropical preschool children. *J. Pediatr. 54*:227–56.

(1967). Approaches to village-level infant feeding. (1) Multimixes as weaning foods. *J. Trop. Pediatr. 13*:46–8.

Jelliffe, D. B., & Jelliffe, E. F. P. (1978). *Human milk in the modern world.* Oxford: Oxford University Press.

Jevons, F. R. (1962). Boerhaave's biochemistry. *Med. Hist. 6*:343–62.

(1963). Boerhaave's teaching in relation to Beccari's identification of gluten as an animal substance. *J. Hist. Med. 18*:174–5.

Johnston, V. J. (1985). *Diet in workhouses and prisons, 1835–1895.* New York: Garland.

Joint FAO/WHO Ad Hoc Expert Committee (1973). *Energy and protein requirements* (WHO Tech. Rep. Ser. No. 522). Geneva: World Health Organization.

Joint FAO/WHO Expert Committee on Nutrition (1950). *Report on the first session* (WHO Tech. Rep. Ser. No. 16). Geneva: World Health Organization.

(1951). *Report on the second session* (WHO Tech. Rep. Ser. No. 44). Geneva: World Health Organization.

(1953). *Report on the third session* (WHO Tech. Rep. Ser. No. 72). Geneva: World Health Organization.

(1955). *Report on the fourth session* (WHO Tech. Rep. Ser. No. 97). Geneva: World Health Organization.

(1958). *Fifth report* (WHO Tech. Rep. Ser. No. 149). Geneva: World Health Organization.

(1962). *Sixth report* (WHO Tech. Rep. Ser. No. 245). Geneva: World Health Organization.

(1967). *Seventh report* (WHO Tech. Rep. Ser. No. 377). Geneva: World Health Organization.

(1971). *Eighth report: Food fortification, protein–calorie malnutrition* (WHO Tech. Rep. Ser. No. 477). Geneva: World Health Organization.

(1976). *Ninth report: Food and nutrition strategies in national development* (WHO Tech. Rep. Ser. No. 584). Geneva: World Health Organization.

Joint FAO/WHO Expert Group (1965). *Protein requirements* (WHO Tech. Rep. Ser. No. 301). Geneva: World Health Organization.

Joint FAO/WHO Informal Gathering of Experts (1975). Energy and protein requirements. *Food. Nutr. 1*(2):11–19.

Kapoor, S. C. (1971). Jean-Baptiste-André Dumas. *Dict. Sci. Biogr. 4*:242–8.

Keill, J. (1728). *Medicina Statica Brittanica* (4th ed., J. Quincy, Ed.). London: Osborn.

Kellogg, J. H. (1877). *A household manual.* Battle Creek, MI: Good Health Publ.

(1893). *Ladies guide in health and disease.* Battle Creek, MI: Modern Medicine Publ.
(1921). *The new dietetics.* Battle Creek, MI: Modern Medicine Publ.
(1922). *Autointoxication or intestinal toxemia.* Battle Creek, MI: Modern Medicine Publ.
(1923). *The natural diet of man.* Battle Creek, MI: Modern Medicine Publ.
Kennedy, E. T., El Lozy, M., & Gershoff, S. (1979). Nutritional need. In J. E. Austin (Ed.), *Global malnutrition and cereal fortification* (pp. 15–34). Cambridge, MA: Ballinger.
Kertesz, Z. I. (1968). Development, formulation and testing of protein mixtures: Past experience. In *Protein foods for the Caribbean* (pp. 71–2). Kingston: Caribbean Food & Nutrition Institute.
Kesselmeyer, J. (1759). *Dissertatio Inauguralis Medica de Quorumdam Vegetabilium Principis Nutriente.* Strasbourg: Kürsner.
Kies, C. V., Shortridge, L., & Reynolds, M. S. (1965). Effect on nitrogen retention of men of varying the total dietary nitrogen with essential amino acids kept constant. *J. Nutr. 85*:260–4.
Kilbourne, E. M. (1992). Eosinophilia – myalgia syndrome: Coming to grips with a new illness. *Epidemiol. Rev. 14*:16–36.
Kohlrausch, O. (1844). *Physiologie und Chemie in ihrer geigenseitigen Stellung beleuchtet durch eine Kritik von Liebigs Thierchemie.* Göttingen: Dieterich.
König, J. (1876). Der Gehalt der menschlichen Nahrungsmittel an Nahrungstoffen im Vergleich zu ihren Preisen. *Z. Biol.* 12:497–512.
Kopple, J. D., & Swendseid, M. E. (1975). Evidence that histidine is an essential amino acid in normal and uremic men. *J. Clin. Invest. 55*:881–91.
Kuhn, T. S. (1970). *The structure of scientific revolutions* (2d ed.). Chicago: University of Chicago Press.
Kutscher, F., & Seemann, J. (1902). Zur Kentniss der Verdauungsvorgänge im Dünndarm, 1. *Z. physiol. Chem. 34*:528–43.
Ladenburg, A. (1900). *Lectures on the history of the development of chemistry since the time of Lavoisier* (L. Dobbin, Trans.) Edinburgh: Alembic Club.
Land, G. (1986). *Adventism in America: A history.* Grand Rapids, MI: Erdmanns.
Landman, J., & Jackson, A. A. (1980). The role of protein deficiency in the aetiology of kwashiorkor. *West Indies Med. J. 29*:229–38.
Langran, M., Moran, B. J., Murphy, J. L., & Jackson, A. A. (1992). Adaptation to a diet low in protein: Effect of complex carbohydrate upon urea kinetics in normal man. *Clin. Sci. 82*:192–8.
Lantigua, R. A., Amatruda, J. M., Biddle, T. L., Forbes, G. B., & Lockwood, D. H. (1980). Cardiac arhythmias associated with a liquid protein diet for the treatment of obesity. *N. Engl. J. Med. 303*:735–8.
Laskowski, N. (1846). Ueber die Proteïn theorie. *Ann. Chem. Pharm. 58*: 129–66.
Lavoisier, A. L. (1778). Rapport sur les blés et farines gâtés. Republished in *Oeuvres de Lavoisier* (Vol. 4, pp. 317–22). Paris: Imprimerie Impériale, 1868.
(1790). *Elements of chemistry* (R. Kerr, Trans.). Edinburgh: Creech (rpt 1965, New York: Dover).
Lawes, J. B. (1853). Agricultural chemistry: Pig feeding. *J. Roy. Agric. Soc. Engl. 14*:459–544.
Lawes, J. B., & Gilbert, J. H. (1859). Experimental inquiry into the composition of some of the animals fed and slaughtered as human food. *Phil. Trans. Roy. Soc. London 149*: 493–680.

Lea, A. S. (1890). A comparative study of artificial and natural digestions. *J. Physiol. 11*:226–63.
Lehmann, C. G. (1842). Untersuchungen über den menschlichen Harn. *J. Prakt. Chem. 27*:257–74.
Leitch, I., & Duckworth, J. (1937). The determination of the protein requirements of man. *Nutr. Abstr. Rev. 7*:257–67.
Lemery, L. (1721). Quatrième mémoire sur les analyses ordinaires des plantes et des animaux. *Mém. Acad. Sci.*, 22–44.
Leroux, A., et al. (1814). Rapport sur un travail de M. d'Arcet, ayant pour objet l'extraction de la gélatine des os, et son application aux différens usages économiques. *Ann. Chim. 92*:300–10.
Lewis, J. S. (1986). *In the family way*. New Brunswick, NJ: Rutgers University Press.
Liebenow, R. C. (1966). Conference introduction. In E. T. Mertz and O. E. Nelson (Eds.), *Proceedings of the High Lysine Corn Conference* (pp. 5–7). Washington DC: Corn Industries Research Foundation.
Liebig, J. (1833). Sur la composition de l'asparamide et de l'acide aspartique. *Ann. Chim. (Ser. 2) 53*:416–21.
(1840). *Organic chemistry in its applications to agriculture and physiology*. (L. Playfair, Trans.) London: Taylor & Walton.
(1841). Ueber die stickstoffhaltigen Nahrungsmittel des Pflanzenreichs. *Ann. Chem. Pharm. 39*:129–60.
(1842a). *Animal chemistry or organic chemistry in its application to physiology and pathology* (W. Gregory, Trans.). Cambridge, MA: Owen.
(1842b). Antwort auf Hrn. Dumas's Rechtfertigung wegen eines Plagiats. *Ann. Chem. Pharm. 41*:351–6.
(1846a). Ueber das Proteïnbioxyd. *Ann. Chem. Pharm. 57*:129–31.
(1846b). *Animal chemistry* (3d ed.; W. Gregory, Trans.) London: Taylor & Walton.
(1847). *Researches on the chemistry of food* (W. Gregory, Trans.). London: Taylor & Walton.
(1867). *Food for infants: A complete substitute for that provided by nature* (2d ed.; E.V. Lersner-Ebersburg, Trans.). London: Walton.
(1870a). The source of muscular power. *Pharm. J. Trans. (Ser. 3) 1*:161–3, 182–5, 201–4, 221–4.
(1870b). *Ueber Gährung, über Quelle de Muskelkraft und Ernährung*. Leipzig: Winter.
(1870c). The process of nutrition. *Pharm. J. Trans. (Ser. 3) 1*:261–3, 281–5.
(1872). Liebig's extract of meat. *The Times* (London), Oct. 1, p. 6.
(1873). Extract of meat. *Pharm. J. Trans. (Ser. 3) 3*:561–3.
Liener, I. E., ed. (1980). *Toxic constituents of plant foodstuffs* (2d ed.). New York: Academic Press.
Loeser, H. A. (1912). Diet of mine natives. *Transvaal Med. J. 7*:152–63.
Lund, C. C., & Levenson, S. M. (1948). Protein nutrition in surgical patients. In M. Sahyun (Ed.), *Proteins and amino acids in nutrition* (pp. 349–63). New York: Reinhold.
Lunn, P. G., Whitehead, R. G., Hay, R. W., & Baker, B. A. (1973). Progressive changes in serum cortisol, insulin and growth hormone concentrations and their relationship to the distorted amino acid pattern during the development of kwashiorkor. *Br. J. Nutr. 29*:399–422.
Lusk, G. (1922). A history of metabolism. In L. F. Barker (Ed.), *Endocrinology and metabolism* (Vol. 3, pp. 3–78). New York: Appleton.

Magendie, F. (1816). Sur les propriétés nutritives des substances qui ne contiennent pas d'azote. *Ann. Chim. (Ser. 2) 3*:66–77, 408–10.
(1831). *An elementary compendium of physiology* (E. Milligan, Trans.), pp. 478–9). Edinburgh: John Carfrae.
(1841). Rapport fait à l'Académie des Sciences au nom de la Commission dite "de la gélatine." *C.R. Acad. Sci. Paris*, 237–83.
Manatt, M. W., & Garcia P. A. (1992). Nitrogen balance: Concepts and techniques. In S. Nissen (Ed.), *Modern methods in protein nutrition and metabolism* (pp. 9–66). San Diego, CA: Academic Press.
Martin, C. J., & Robison, R. (1922). The minimum nitrogen expenditure of man and the biological value of various proteins for human nutrition. *Biochem. J. 16*:407–47.
Mateles, R. I., & Tannenbaum, S. R., Eds. (1968). *Single-cell protein*. Cambridge, MA: MIT Press.
Matthews, D. M. (1991). *Protein absorption*. New York: Wiley-Liss.
Mauron, J. (1980–1). Have single-cell proteins a future? *Nestlé Research News*, 70–9.
(1986). Food, mood and health: The medieval outlook. *Internat. J. Vit. Nutr. Rev.* 56 (suppl. 29):9–26.
Mayer, J. R. (1845). *Die organische Bewegung in ihren Zusammenhange mit dem Stoffwechsel*. Heilbronn: Drecksler.
Mayhew, H. J., & Binny, J. (1862). *The criminal prisons of London and scenes of prison life* (rpt. 1968, London: Cass).
Maynard, L. A. (1962). Wilbur O. Atwater: A biographical sketch. *J. Nutr. 78*:3–9.
Mayow, J. (1674). *Medico-physical works* (trans. 1908). Chicago: University of Chicago Press.
Mazel, J. (1981). *The Beverly Hills Diet*. New York: Macmillan.
McCance, R. A. (1968). The two syndromes. In R. A. McCance & E. M. Widdowson (Eds.), *Calorie deficiencies and protein deficiencies* (pp. 1–4). London: Churchill.
McCay, D. (1912). *The protein element in nutrition*. London: Edward Arnold.
McCollum, E. V. (1957). *A history of nutrition*. Boston: Houghton Mifflin.
McCosh, F. W. J. (1984). *Boussingault, chemist and agriculturalist*. Dordrecht: Reidel.
McCoy, R. H., Meyer, C. E., & Rose, W. C. (1935). Feeding experiments with mixtures of highly purified amino acids, 8. *J. Biol. Chem. 112*:283–302.
McKie, D. (1944). Wöhler's "synthetic" urea and the rejection of vitalism: A chemical legend. *Nature 153*:608–10.
(1965). *Introduction to Lavoisier's "Elements of Chemistry" (1790)*. New York: Dover.
McLaren, D. S. (1966). A fresh look at protein–calorie malnutrition. *Lancet 2*:485–8.
(1967). A criticism of the rationale for the development of vegetable food mixtures for the prevention of protein–calorie malnutrition in young children. *Proc. 7th Internat. Congr. Nutr., Hamburg 3*:149.
(1968). Vitamin deficiencies complicating the several forms of protein–calorie malnutrition, with special reference to vitamin A. In R. A. McCance & E. M. Widdowson (Eds.), *Calorie deficiencies and protein deficiencies* (pp. 191–9). London: Churchill.
(1972). Expert committees. *Nature 237*:119.

(1973). Energy–protein malnutrition. *Lancet 1*:994.
(1974). The great protein fiasco. *Lancet 2*:93–6, 107–9.
(1978). Nutrition planning day dreams at the United Nations. *Am. J. Clin. Nutr. 31*:1295–9.
Medsger, T. A. (1990). Tryptophan-induced eosinophilia-myalgia syndrome. *N. Engl. J. Med. 322*:926–8.
Meguid, M. M., Matthews, D. E., Bier, D. M., Meredith, C. N., Soeldner, J. S., & Young, V. R. (1986). Leucine kinetics at graded leucine intakes in young men. *Am. J. Clin. Nutr.* 43:770–80.
Mendel, L. B. (1923). *Nutrition: The chemistry of life*. New Haven, CT: Yale University Press.
Mendelsohn, E. (1964). *Heat and life*. Cambridge, MA: Harvard University Press.
Mertz, E. T. (1992). *Quality protein maize*. St. Paul, MN: Am. Assoc. Cereal Chemists.
Mertz, E. T., Bates, L. S., & Nelson, O. E. (1964). Mutant gene that changes protein composition and increases lysine content of maize endosperm. *Science 145*:279–80
Meyer, E. V. (1891). *A history of chemistry from the earliest times to the present day*. London: Macmillan.
Michaud, L. (1909). Beitrag zur Kenntnis des physiologischen Eiweissminimum. *Z. physiol. Chem. 59*:405–91.
Millward, D. J. (1985). Human protein requirements: The physiological significance of changes in the rate of whole-body protein turnover. In J. S. Garrow & D. Halliday (Eds.), *Substrate and energy metabolism* (pp. 135–44). London: Libbey.
Millward, D. J., Price, G. M., Pacy, P. J. H., & Halliday, D. (1991). Whole-body protein and amino acid turnover in man: What can we measure with confidence? *Proc. Nutr. Soc. 50*:197–216.
Mitchell, H. H. (1924a). A method of determining the biological value of protein. *J. Biol. Chem. 58*:873–903.
(1924b). The biological value of proteins at different levels of intake. *J. Biol. Chem. 58*:905–22.
(1924c). The nutritive value of proteins. *Physiol. Rev.* 4:424–78.
(1942). The metabolism of protein and amino acids. *Ann. Rev. Biochem. 11*:257–82.
Mitchell, H. H., & Block, R. J. (1946). Some relationships between the amino acid contents of proteins and their nutritive values for the rat. *J. Biol. Chem. 163*:599–620.
Mitchell, H. H., Burroughs, W., & Beadles, J. R. (1936). The significance and accuracy of biological values of proteins computed from nitrogen metabolism data. *J. Nutr. 11*:257–74.
Mitchell, H. H., & Carman, G. G. (1924). The biological value for maintenance and growth of whole wheat, eggs, and pork. *J. Biol. Chem. 60*:613–20.
(1926). The biological value of the nitrogen of mixtures of patent white flour and animal foods. *J. Biol. Chem. 68*:183–215.
Money, J. (1985). *The destroying angel*. Buffalo, NY: Prometheus.
Moo-Young, M., & Gregory, K. F., eds. (1986). *Microbial biomass proteins*. London: Elsevier.
Moore, B. (1898). Chemistry of the digestive processes. In E. A. Schäfer (Ed.), *Textbook of physiology* (Vol. 1, pp. 312–474). Edinburgh: Pentland.

Mortimer, C. (1745). The natural heat of animals. *Phil. Trans. Roy. Soc. 43*:473–80.
Mueller, J. H. (1923). A new sulfur-containing amino-acid isolated from the hydrolytic products of protein. *J. Biol. Chem. 56*:57–69.
Mulder, G. J. (1839). Über die Zusammensetzung einiger thierischen Substanzen. *J. Prakt. Chem. 16*:129–52.
(1846). *Liebig's question to Mulder* (P. F. H. Fromberg, Trans.). London: Blackwood.
Müller, M. (1906). Untersuchungen über die bisher beobachtete eiweisssparende Wirkung des Asparagins bei der Ernährung. *Arch. ges. Physiol. 112*:245–91.
Munday, P. (1990). Social climbing through chemistry: Justus Liebig's rise from the *Niederer Mittelstand* to the *Bildungsbürgertum*. *Ambix 37*:1–6.
(1991). Liebig's metamorphosis: From organic chemistry to the chemistry of agriculture. *Ambix 38*:135–53.
Munk, I. (1893). Ueber die Folgen einer ausreichenden, aber eiweissarmen Nahrung. *Virchow's Arch. path. Anat. Physiol. 132*:91–157.
Munro, H. N. (1964). Regulation of protein metabolism. In H. N. Munro & J. B. Allison (Eds.), *Mammalian protein metabolism* (Vol. 1, pp. 382–481). New York: Academic Press.
(1985). Historical perspectives on protein requirements: Objectives for the future. In K. L. Blaxter & J. C. Waterlow (Eds.), *Nutritional adaptation in man* (pp. 155–67). London: John Libbey.
Naismith, D. J. (1973). Kwashiorkor in Western Nigeria: A study of traditional weaning foods, with particular reference to energy and linoleic acid. *Br. J. Nutr. 30*:567–76.
National Nutrition Research Institute (1959). *Food enrichment in South Africa*. Pretoria: South African Council for Scientific and Industrial Research.
National Research Council (1989). *Recommended dietary allowances* (10th ed.). Washington, DC: National Academy Press.
Naylor, M. V. (1942). Sylvester Graham, 1794–1851. *Ann. Med. Hist. (Ser. 3) 4*:236–40.
Needham, D. M. (1971). *Machina carnis: The biochemistry of muscular contraction in its historical development*. Cambridge: Cambridge University Press.
Newton, J. F. (1811). *The return to nature or a defence of the vegetable regimen*. London: Cadell & Davies.
Nicol, B. M. (1962). The utilization of protein-rich foods in the prevention of protein–calorie deficiency diseases. In G. Blix (Ed.), *Mild–moderate forms of protein–calorie malnutrition* (pp. 152–8). Uppsala: Swedish Nutrition Foundation.
(1971). Protein and calorie concentration. *Nutr. Rev. 29*:83–8.
Nissenbaum, S. (1980). *Sex, diet, and debility in Jacksonian America*. Westport, CT: Greenwood Press.
Nitti, F. S. (1896). The food and labour power of nations. *Econ. J. 6*:30–63.
Nutrition Expert Group (1968). *Recommended daily allowances of nutrients and balanced diets*. Hyderabad: Indian Council of Medical Research.
ODA Advisory Committee on Protein. (1975). *British aid and the relief of malnutrition* (Overseas Development Paper No. 2). London: H.M. Stationery Office.
Oddoye, E. A., & Margen, S. (1979). Nitrogen balance studies in humans: Long-term effect of high nitrogen intake on nitrogen accretion. *J. Nutr. 109*:363–77.

O'Hara May, J. (1977). *Elizabethan dyetary of health*. Lawrence, KA: Coronado Press.
Orr, E. (1972). *The use of protein-rich foods for the relief of malnutrition in developing countries: An analysis of experience*. London: Tropical Products Institute.
(1977). The contribution of new food mixtures to the relief of malnutrition: A second look. *Food Nutr.* 3(2):2–10.
Orr, E., & Adair, D. (1967). *The production of protein foods and concentrates from oilseeds*. London: Tropical Products Institute.
Orr, J. B. (1937). *Food, health and income* (2d ed.). London: Macmillan.
Osborne, T. B., & Jones, D. B. (1910). A consideration of the sources of loss in analyzing the products of protein hydrolysis. *Am. J. Physiol.* 26:305–28.
Osborne, T. B., & Mendel, L. B. (1911). *Feeding experiments with isolated food-substances* (Publ. No. 156). Washington, DC: Carnegie Institution of Washington.
(1914). Amino-acids in nutrition and growth. *J. Biol. Chem.* 17:325–49.
Osborne, T. B., Mendel, L. B. & Ferry, E. L. (1919). A method of expressing numerically the growth-promoting value of proteins. *J. Biol. Chem.* 37:223–9.
Pagel, W. (1956). Van Helmont's ideas on gastric digestion and the gastric acid. *Bull. Hist. Med.* 30:524–36.
Panel of Experts on the Protein Problem Confronting Developing Countries (1971). *Strategy statement on action to avert the protein crisis in the developing countries*. New York: United Nations.
Paris, J. A. (1826). *A treatise on diet*. London: Underwood.
Pariser, E. R., Wallerstein, M. B., Corkery, C. J., & Brown, N. L. (1978). *Fish protein concentrate: Panacea for protein malnutrition?* Cambridge, MA: MIT Press.
Parkes, E. A. (1871). Further experiments on the effect of diet and exercise on the elimination of nitrogen. *Proc. Roy. Soc. London* 19:349–61.
Parmentier, A. (1773). *Examen chymique des pommes de terre*. Paris: Didet.
Parpia, H. A. B. (1969). Protein foods of India based on cereals, legumes, and oilseeds. In M. Milner (Ed.), *Protein-enriched cereal foods for world needs* (pp. 129–39). St. Paul, MN: American Association of Cereal Chemists.
Partington, J. R. (1961–4). *A history of chemistry* (Vols. 2–4). London: Macmillan.
Paton, D. N. (1895). Muscular energy: The present state of our knowledge in regard to its source. *Edinburgh Med. J.* 40:1081–91.
Pereira, J. (1843). *A treatise on food and diet*. London: Longmans.
Peterson, G. E. (1964). *The New England college in the age of the university*. Amherst, MA: Amherst College Press.
Petit, et al. (1831). Rapport des médecins, chirurgiens et pharmaciens de l'Hôtel-Dieu. *C.R. Acad. Sci. Paris,* 13 (1841):286–95.
Pettenkoffer, M., & Voit. C. (1866). Untersuchungen über den Stoffverbrauch des normalen Menschen. *Z. Biol.* 2:459–573.
Pitcairn, A. (1718). *The philosophical and mathematical elements of physick*. London: Bell & Osborn.
Playfair, L. (1843). Second lecture on the rearing and feeding of cattle. *J. Roy. Agric. Soc. England* 4:237–66.
(1853). On the food of man under different conditions of age and employment. *Proc. Roy. Inst. Great Brit.* 1:313–16.
Poleman, T. T. (1977). World food: Myth and reality. *World Develop.* 5:383–94.

Poppel, J. (1869). Erfährungen über die Liebig'sche Malzsuppe. *Berliner Klin. Woch. 6*:494–6.

Prahlad Rao, N., Singh, D., & Swaminathan, M. C. (1969). Nutritional status of pre-school children of rural communities near Hyderabad City. *Ind. J. Med. Res. 57*:2132–46.

Pretorius, P. J., & Smit, Z. M. (1968). The effects of various skimmed milk formulae on the diarrhea, nitrogen retention and initiation of cure in kwashiorkor. *J. Trop. Pediatr.* 4:50–60.

Protein Advisory Group (1970). Guideline for preparation of edible cottonseed protein concentrate. Republished in *The PAG compendium (1975)* (Vol. C1, pp. 1137–40). New York: Worldmark Press.

(1973). PAG statement (no.20) on the "protein problem." *PAG Bull.* 3 (1): 4–10.

Proust, L. J. (1802). Essai sur la fécule des plantes vertes. *J. Phys. Chim. Hist. nat. 56*:97–113.

Prout, W. (1819). Proprietés chimiques et composition de quelques-uns des principes immédiats de l'urine. *Ann. Chim. Phys. (Ser. 2) 10*:369–88.

(1845). *Chemistry, meteorology and the function of digestion* (3d ed.). London: Churchill.

Quesnay, F. (1747). *Essai physique sur l'oeconomie animale* (3 vols.). Paris: Cavelier.

Rand, W. M., Uauy, R., & Scrimshaw, N. S., eds. (1984). Protein–energy requirement studies in developing countries: Results of international research. *Food Nutr. Bull., Suppl No. 10.*

Rao, K. S., Swaminathan, M. C., Swarup, S., & Patwardhan, V. N. (1959). Protein malnutrition in South India. *Bull. World Health Org. 20*:603–39.

Rappaport, R. (1960). G.-F. Rouelle: An eighteenth century chemist and teacher. *Chymia 6*:68–101.

Ratner, S. (1979). The dynamic state of body proteins. *Ann. N. Y. Acad. Sci. 325*:189–209.

Reeds, P. J. (1990). Amino acid needs and protein scoring patterns. *Proc. Nutr. Soc. 49*:489–97.

Rhodes, K. (1957). Two types of liver disease in Jamaican children, 3. *West Ind. Med. J. 6*:145–78.

Ritson, J. (1802). *An essay on abstinence from animal food as a moral duty.* London: Richard Phillips.

Ritthausen, H. (1872). *Die Eiweisskörper der Getreidarten, Hülsenfrüchte und Ölsamen.* Bonn: Max Cohen & Sohn.

Rivers, J. P. W. (1979). The profession of nutrition: An historical perspective. *Proc. Nutr. Soc. 38*:225–31.

Roe, D. A. (1981). William Cumming Rose: A biographical sketch. *J. Nutr. 111*:1313–20.

Roels, O. A. (1972). Fish protein concentrate: History and trends in production. In *Production of fish-protein concentrate* (Part 2). New York: United Nations.

Rolleston, H. (1940). Walter Charleton, D.M., F.R.C.P., F.R.S. *Bull. Hist. Med. 8*:403–16.

Rose, W. C. (1931). Feeding experiments with mixtures of highly purified amino acid: 1. The inadequacy of diets containing nineteen amino acids. *J. Biol. Chem. 94*:155–65.

(1936). The significance of the amino acids in nutrition. In *The Harvey Lectures* (Vol. 30, pp. 49–65). Baltimore: Wilkins.

(1949). Amino acid requirements of man. *Fed. Proc. 8*:546–52.

(1957). The amino acid requirements of adult man. *Nutr. Abstr. Rev.* *27*:631–47.
Rose, W. C., Oesterling, M. J., & Womack, M. (1948). Comparative growth on diets containing ten and nineteen amino acids, with further observations upon the role of glutamic and aspartic acid. *J. Biol. Chem.* *176*:753–62.
Rose, W. C., & Wixom, R. L. (1955). The amino acid requirements of man: 16. The role of nitrogen intake. *J. Biol. Chem.* *217*:997–1004.
Rosen, E., Buchanan, N., & Hansen, J. D. L. (1974). Evolution of kwashiorkor and marasmus. *Lancet* *2*:458.
Rosen, G. D. (1958). Groundnuts (peanuts) and groundnut meal. In A. A. Altschul (Ed.), *Processed plant protein foodstuffs* (pp. 419–68). New York: Academic Press.
Rosenheim, T. (1893). Weiterer Untersuchungen über die Schädlichkeit eiweissarmer Nahrung. *Pflüger's Arch. Ges. Physiol.* *54*:61–71.
Rossiter, M. W. (1975). *The emergence of agricultural science: Justus Liebig and the Americans*. New Haven, CT: Yale University Press.
Rothschuh, K. E. (1971). Adolf Eugen Fick. *Dict. Sci. Biogr.* *4*:614–17.
(1975). Max Rubner. *Dict. Sci. Biogr.* *11*:585–6.
Rouelle, H. M. (1773a). Expériences. *J. Méd. Chir. Pharm.* *39*:250–66.
(1773b). Observations sur les fécules ou parties vertes des plantes et sur la matière glutineuse ou végéto animale. *J. Méd. Chir. Pharm.* *40*:59–67.
(1773c). Sur l'urin humain, et sur celles de vache et cheval. *J. Méd.* *40*:451–68.
Rubner, M. (1879). Ueber die Ausnützung einer Nahrungsmittel im Darmcanale des Menschen. *Z. Biol.* *15*:115–202.
(1908). Ernährungsvorgänge beim Wachstum des Kindes. *Arch. Hyg.* *66*:81–126.
(1911). Verluste und Wiedererneuerung im Lebensprozess. *Arch. Anat. Physiol.*, 39–60.
Rumford, B. (1800). *Essays, political, economical, and philosophical*. London: Codell & Davies.
Rush, B. (1812). *Medical inquiries and observations upon the diseases of the mind*. Philadelphia: Kimber & Richardson.
Rutishauser, I. H. E., & Whitehead, R. G. (1972). Energy intake and expenditure in 1–3 year old Ugandan children living in a rural environment. *Br. J. Nutr.* *28*:145–52.
Salkowski, E. (1880). Weitere Beiträge zur Theorie der Harnstoffbildung. *Z. physiol. Chem.* *4*:100–32.
Sanchez, L., Calvo, M., & Brock, J. H. (1992). Biological role of lactoferrin. *Arch. Dis. Child.* *67*:657–61.
Santorio, S. (1614). Insensible perspiration. Translation in J. F. Fulton & L. G. Wilson (Eds.), *Selected readings in the history of physiology* (1966) (pp. 162–3). Springfield, IL: Thomas.
Schatan, J. (1977). Termination of the PAG. *PAG Bulletin* 7 (3–4): 1–20.
Scherer, J. (1841). Chemisch-physiologische Untersuchungen. *Ann. Chem. Pharm.* *40*:1–64.
Schoenheimer, R. (1942). *The dynamic state of body constituents*. Cambridge, MA: Harvard University Press.
Schoenheimer, R., Ratner, S., & Rittenberg, D. (1939). The metabolic activity of body proteins investigated with *l*(–)-leucine containing two isotopes. *J. Biol. Chem.* *130*:703–32.
Scrimshaw, N. S. (1973). The "protein problem." *Nutrition Notes (American Institute of Nutrition)* Sept.:3–4.

(1976). Shattuck lecture: Strengths and weaknesses of the committee approach. *N. Engl. J. Med.* 294:136–42, 198–203.
(1980). The background and history of Incaparina. *Food Nutr. Bull.* 2:1–2.
Scrimshaw, N. S., Behar, M., Wilson, D., Viteri, F., Arroyave, G., & Bressani, R. (1961). All-vegetable protein mixtures for human feeding: 5. Clinical trials with INCAP mixtures 8 and 9 and with corn and beans. *Am. J. Clin. Nutr.* 9:196–205.
Scrimshaw, N. S., & Dillon, J.-C. (1979). Allergic responses to some single-cell proteins in human subjects. In S. Garattini, S. Paglialunga, & N. S. Scrimshaw (Eds.), *Single-cell protein: Safety for human and animal feeding* (pp. 171–8). Oxford: Pergamon.
Scrimshaw, N. S., Hussein, M. A., Murray, E., Rand, W. M., & Young, V. R. (1972). Protein requirements of man: Variations in obligatory urinary and fecal nitrogen losses in young men. *J. Nutr.* 102:1595–604.
Scrimshaw, N. S., & Young, V. R. (1989). Adaptation to low protein and energy intakes. *Human Org.* 48:20–30.
Sebrell, W. H., et al., eds. (1961). *Progress in meeting protein needs of infants and preschool children* (Publ. No. 843). Washington, DC: National Research Council.
Seguin, A., & Lavoisier, A. L. (1789). Premier mémoire sur la respiration des animaux. *Mém. Acad. Sci.*, 566–84.
Sharp, N. A. D. (1935). A note on a nutritional disease of childhood. *Trans. Roy. Soc. Trop. Med. Hyg.* 28:411–12.
Shaw, R. L. (1969). Incaparina in Central America. In M. Milner (Ed.), *Protein-enriched cereal foods for world needs* (pp. 320–33). St. Paul, MN: American Association of Cereal Chemists.
Shelley, P. B. (1813). *A vindication of natural diet.* London: Callow.
Shenstone, W. A. (1901). *Justus von Liebig: His life and work.* London: Cassell.
Sherman, H. C. (1920). Protein requirement of maintenance in man and the nutritive efficiency of bread protein. *J. Biol. Chem.* 41:97–109.
Smeaton, W. A. (1962). *Fourcroy: Chemist and revolutionary.* Cambridge: Heffer.
Smith, A. H. (1956). Lafayette Benedict Mendel. *J. Nutr.* 60:3–12.
Smith, A. K., & Circle, S. J. (1972). *Soybean: Chemistry and technology.* Westport, CT: Avi.
Smith, E. (1857a). On the principles involved in a scheme of prison dietary. *Trans. Nat. Assoc. Promotion Soc. Sci.*, 293–306.
(1857b). The influence of the labour of the treadwheel over respiration and pulsation. *Med. Times Gaz.* 14:601–3.
(1859). Experimental inquiries into the chemical and other phenomena of respiration and their modification by various physical agencies. *Phil. Trans. Roy. Soc. London* 149:681–714.
(1862). On the elimination of urea and urinary water. *Phil. Trans. Roy. Soc. London* 151:747–834.
(1864). Report on the food of the poorer laboring classes in England. *British Parliamentary Papers for 1864* 28:216–329.
(1874). *Foods* (3d ed.). London: King.
Smuts, D. B. (1935). The relation between the basal metabolism and the endogenous nitrogen metabolism, with particular reference to the estimation of the maintenance requirement of protein. *J. Nutr.* 9:403–33.
Snelders, H. A. M. (1982). The Mulder–Liebig controversy elucidated by their correspondence. *Janus* 69:199–221.

Snell, E. E. (1979). Lactic acid bacteria and identification of B-vitamins: Some historical notes, 1937–40. *Fed. Proc. 38*:2690–3.

Sokolow, J. A. (1983). *Eros and modernization: Sylvester Graham, health reform, and the origins of Victorian sexuality in America.* London: Associated Universities Press.

Stannus, H. S. (1934). A nutritional disease of childhood associated with a maize diet and pellagra. *Arch. Dis. Child. 9*:115–18.

(1935). Kwashiorkor. *Lancet 2*:1207–8.

(1936). Pellagra and pellagra-like conditions in warm climates. *Trop. Dis. Bull. 33*:729–41, 815–25, 885–901.

Stroup, A. (1990). *A company of scientists: Botany, patronage and community at the seventeenth-century Parisian Royal Academy of Sciences.* Berkeley and Los Angeles: University of California Press.

Sukhatme, P. V. (1966). The world's food supplies. *J. Roy. Stat. Soc. (Ser. A) 129*:222–43.

(1970a). Incidence of protein deficiency in relation to different diets in India. *Br. J. Nutr. 24*:477–87.

(1970b). Size and nature of the protein gap. *Nutr. Rev. 28*:223–6.

Takaki, K. (1885). On the cause and prevention of Kak'ke. *Sei-i-kai Med. J. 4*:29–37.

Tannenbaum, S. R., & Wang, D. I. C., Eds. (1975). *Single-cell protein (Vol. 2).* Cambridge, MA: MIT Press.

Teich, M. (1965). On the historical foundations of modern biochemistry. *Clio Medica 1*:41–57.

(1992). *A documentary history of biochemistry, 1770–1940.* Rutherford, NJ: Fairleigh Dickinson University Press.

Temkin, O. (1960). Nutrition from classical antiquity to the baroque. In I. Goldston (Ed.), *Human nutrition, historic and scientific* (pp. 78–97). New York: International Universities Press.

Thomas, J. L. (1965). Romantic reform in America, 1815–1865. *Am. Quart. 17*:656–81.

Thomas, K. (1909). Über die biologische Wertigkeit der Stickstoffsubstanzen in verschiedenen Nahrungsmitteln. *Arch. Physiol. Leipzig,* 219–302.

Thouvenel, P. (1806). *Mélanges d' histoire naturelle, de physique et de chimie* (3 vols.). Paris: Bertrand & Colnet.

Tomlinson, M. H. (1978). "Not an instrument of punishment": Prison diet in the mid-nineteenth century. *J. Consum. Stud. Home Econ. 2*:15–26.

Torun, B., Pineda, O., Viteri, F. E., & Arroyave, G. (1981). Use of amino acid composition data to predict nutritive value for children with specific reference to new estimates of their essential amino acid requirements. In C. E. Bodwell, N. S. Adkins & D. T. Hopkins (Eds.), *Protein quality in humans* (pp. 374–93). Westport, CT: Avi.

Torun, B., Young, V. R., & Rand, W. D., eds. (1981). Protein–energy requirements of developing countries: Evaluation of new data. *Food Nutr. Bull.,* Suppl No. 5.

Toulmin, S., & Goodfield, J. (1962). *The architecture of matter.* New York: Harper & Row.

Traube, M. (1861). Ueber die Beziehung der Respiration zur Musckelthätigkeit und die Bedeutung der Respiration uberhaupt. *Arch Path. Anat. 21*:386–414.

Trémolieres, J. (1975). A history of dietetics. *Prog. Food Nutr. Sci. 1*:65–114.

Trowell, H. C. (1940). Infantile pellagra. *Trans. Roy. Soc. Trop. Med. Hyg. 33*:389–404.

(1944). Malnutrition in Bantu of Central Africa: Syndrome of malignant malnutrition. *Clin. Proc. 3*:381–401.
(1946). The kwashiorkor syndrome. *Trans. Roy. Soc. Trop. Med. Hyg. 39*:272.
(1949). Malignant malnutrition. *Trans. Roy. Soc. Trop. Med. Hyg.* 42:417–33.
(1982). The beginning of the kwashiorkor story in Africa. Preface to H. C. Trowell, J. N. P. Davies & R. F. A. Dean, *Kwashiorkor* (pp. xxi–xxviii). New York: Academic Press.
Trowell, H. C., Davies, J. N. P., & Dean, R. F. A. (1954). *Kwashiorkor*. London: Edward Arnold (republished with additional prefaces in 1982, New York: Academic Press).
Trowell, H. C., & Muwazi, E. M. K. (1945). A contribution to the study of malnutrition in Central Africa: A syndrome of malignant malnutrition. *Trans. Roy. Soc. Trop. Med. Hyg. 39*:229–43.
Underhill, F. P., & Mendel, L. B. (1928). A dietary deficiency canine disease: Further experiments. *Am. J. Physiol. 83*:589–633.
United Nations (1975). *Report of the World Food Conference, Rome, 1974*. New York: United Nations..
U.S. Department of Agriculture (1983). *Nationwide Food Consumption Survey, 1977–78 (Report No. H-6)*. Washington, DC: U.S. Government Printing Office.
(1985). *Nationwide Food Consumption Survey. 1977–78*, (Report No. H-11). Washington, DC: U.S. Government Printing Office.
Vaghefi, S. B., Makdani, D. D., & Mickelsen, O. (1974). Lysine supplementation of wheat proteins: A review. *Am. J. Clin. Nutr. 27*:1231–43.
Van-Bochaute, _ (1786). Mémoire sur l'origine et de la nature de la substance animale. *Obs. physique 28*:109–15.
Vauquelin, L. N., & Fourcroy, A. N. (1806). Memoir upon the germination and fermentation of grains and farinaceous substances. *Phil. Mag. 25*:176–82.
Venel, G.-F. (1755). Essai sur l'analyse des végétaux: Premier mémoire. *Mém. Acad. Sci.* 2 319–32.
Vickery, H. B. (1945). Russell Henry Chittenden, 1865–1943. *Nat. Acad. Sci. Biogr. Mem. 24*:59–104.
(1950). The origin of the word protein. *Yale J. Biol. Med. 22*:387–93.
(1956). Thomas Burr Osborne. *J. Nutr. 59*:3–26.
Vickery, H. B., & Osborne, T. B. (1928). A review of hypotheses of the structure of proteins. *Physiol. Rev. 8*:393–446.
Vickery, H. B., & Schmidt, C. L. A. (1931). The history of the discovery of the amino acids. *Chem. Rev. 9*:169–318.
Voit, C. (1860). *Untersuchungen über den Einfluss des Kochsalzes, des Kaffee's und der Muskelbewegungen auf den Stoffwechsel*. Munich: Cotta.
(1867). Der Eiweissumsatz bei Ernährung mit reinem Fleisch. *Z. Biol. 3*:1–85.
(1869). Ueber den Eiweissumsatz bei Zufuhr von Eiweiss und Fett, und über die Bedeutung des Fettes für die Ernährung. *Z. Biol. 5*:329–68.
(1870). Ueber der Lehre von der Quelle der Muskelkraft und einiger Theile der Ernährung seit 25 Jahren. *Z. Biol. 6*:303–401.
(1876). Ueber die Kost in öffentlichen Anstalten. *Z. Biol. 12*:1–59
(1881). Physiologie des allgemeines Stoffwechsels und der Ernährung. In L. Hermann (Ed.), *Handbuch der Physiologie* (Vol. 6, Part 1, pp. 1–575). Leipzig: Vogel.
(1889). Ueber die Kost eines Vegetariers. *Z. Biol. 25*:232–88.
Waslien, C. I., Calloway, D. H., & Margen, S. (1969). Human intolerance to bacteria as food. *Nature 221*:84–5.

Waterlow, J. C. (1948a). Comment on kwashiorkor. *Trop. Dis. Bull. 45:* 637–8.

(1948b). *Fatty liver disease in infants in the British West Indies* (Medical Research Council Spec. Rep. Ser. No. 263). London: H.M. Stationery Office.

(1961). The rate of recovery of malnourished infants in relation to the protein and calorie levels of the diet. *J. Trop. Ped. 7:*16–22.

(1974). Evolution of kwashiorkor and marasmus. *Lancet 2:*712.

(1984). Kwashiorkor revisited: The pathogenesis of oedema in kwashiorkor and its significance. *Trans. Roy. Soc. Trop. Med. Hyg. 78:*436–41.

(1985). What do we mean by adaptation? In K. L. Blaxter & J. C. Waterlow (Eds.), *Nutritional adaptation in man* (pp. 1–10). London: John Libbey.

(1990). Protein requirements of infants: an operational assessment. *Proc. Nutr. Soc. 49:*499–506.

(1992). *Protein energy malnutrition*. London: Edward Arnold.

Waterlow, J.C., ed. (1955). *Protein malnutrition* (FAO/WHO/FAO Nutrition Meetings Rep. No. 12). Geneva: United Nations.

Waterlow, J. C., & Payne, P. R. (1975). The protein gap. *Nature, London 258:*113–17.

Waterlow, J. C., & Stephen, J. M. L., Eds. (1957). *Human protein requirements and their fulfilment in practice* (FAO Nutritional Meeting Rep. Ser. No. 12). Rome: United Nations.

Waterlow, J. C., & Vergara, A. (1956). *Protein malnutrition in Brazil.* (FAO Nutritional Studies No. 14). Rome: United Nations.

Waterlow, J. C., & Wills, V. G. (1960). Balance studies in malnourished Jamaican infants: 1. Absorption and retention of nitrogen and phosphorus. *Br. J. Nutr.* 14:183–98.

Welbourn, H. (1955). The danger period during weaning. *J. Trop. Ped. 1:*34–46, 98–111, 161–73.

Weller, L. A., Calloway, D. H., & Margen, S. (1971). Nitrogen balance of men fed amino acid mixtures based on Rose's requirements, egg white protein and serum free amino acid patterns. *J. Nutr. 101:*1499–508.

Wharton, B. A. (1968). Difficulties in the initial treatment of kwashiorkor. In R. A. McCance & E. M. Widdowson (Eds.), *Calorie deficiencies and protein deficiencies* (pp. 147–53). London: Churchill.

Wharton, B. A., Howells, G. R., & McCance, R. A. (1967). Cardiac failure in kwashiorkor. *Lancet 2:*384–7.

Whitehead, R. G. (1969). Factors which may affect the biochemical response to protein–calorie malnutrition. In A. V. Muralt (Ed.), *Protein–calorie malnutrition.* (pp. 38–47). Berlin:Springer.

(1973). The protein needs of malnourished children. In J. W. G. Porter & B. A. Rolls (Eds.), *Proteins in human nutrition* (pp. 103–17). London: Academic Press.

Whitehead, R. G., & Alleyne, G. A. O. (1972). Pathophysiological factors of importance in protein–calorie malnutrition. *Br. Med. Bull. 28:*72–8.

Whitehead, R. G., Coward, W. A., Lunn, P. G., & Rutishauser, I. (1977). A comparison of the pathogenesis of protein–energy malnutrition in Uganda and the Gambia. *Trans. Roy. Soc. Trop. Med. Hyg. 71:*189–98.

Whorton, J. C. (1982). *Crusaders for fitness: The history of American health reformers.* Princeton, NJ: Princeton University Press.

Widdowson, E. M., & McCance, R. A. (1954). *Studies on the nutritive value of bread and on the effect of variations in the extraction rate of flour on the growth of undernourished children* (Medical Research Council Special Rep. Ser. No. 287). London: H.M. Stationery Office.

Willcock, E. G., & Hopkins, F. G. (1906). The importance of individual amino-acids in metabolism. *J. Physiol.* *35*:88–102.

Williams, C. D. (1933). A nutritional disease of childhood associated with a maize diet. *Arch. Dis Child.* *8*:423–33.

(1935). Kwashiorkor, a nutritional disease of children associated with a maize diet. *Lancet* 2:1151–2.

(1938). Child health in Gold Coast. *Lancet* *1*:97–102.

(1940). What is pellagra in children? *Trans. Roy. Soc. Trop. Med. Hyg.* *34*:85–90.

(1973). The story of kwashiorkor. *Nutr. Rev.* *31*:334–40.

Williams, R. R. (1961). *Toward the conquest of beriberi.* Cambridge, MA: Harvard University Press.

Willis, T. (1684). *A medical-philosophical discourse on fermentation.* London: Dring, Harper & Leigh.

Wise, R. P. (1980). The case of Incaparina in Guatemala. *Food Nutr. Bull.* *2*:3–8.

Wöhler, F. (1828). On the artificial production of urea. *Quart. J. Sci. Art.* *25*:491–2.

Yates, F. E. (1983). Contribution of statistics to the ethics of science. *Am. J. Physiol.* *244*:R3–5.

Young, V. R. (1986). Nutritional balance studies: Indicators of human requirement or of adaptive mechanisms. *J. Nutr.* *116*:700–3.

Young, V. R., Bier, D. M., & Pellet, P. L. (1989). A theoretical basis for increasing current estimates of the amino acid requirements in adult man, with experimental support. *Am. J. Clin. Nutr.* *50*:80–92.

Young, V. R., Gucalp, C., Rand, W. M., Matthews, D. E., & Bier, D. M. (1987). Leucine kinetics during three weeks at submaintenance-to-maintenance intakes of leucine in men: Adaptation and accommodation. *Hum. Nutr. Clin. Nutr.* *41C*:1–18.

Young, V. R., & Pellet, P. L. (1988). How to evaluate dietary protein. In C. A. Barth & E. Schlimmer (Eds.), *Milk proteins: Nutritional, clinical, functional and technological aspects* (pp. 7–36). New York: Springer.

Young, V. R., Puig, M., Queiroz, E., Scrimshaw, N. S., & Rand, W. M. (1984). Evaluation of the protein quality of an isolated soy protein in young men: Relative nitrogen requirements and effect of methionine supplementation. *Am. J. Clin. Nutr.* *39*:16–24.

Zello, G. A., Pencharz, P. B., & Ball, R. O. (1993). Dietary lysine requirement of young adult males determined by oxidation of L-[1-^{13}C]phenylalanine. *Am. J. Physiol.* *264*:E677–85.

Zezulka, A. Y., & Calloway, D. H. (1976). Nitrogen retention in men fed varying levels of amino acids from soy protein with or without added L-methionine. *J. Nutr.* *106*:212–21.

Zuntz, N. (1897). The metabolism of nutrients in the animal body and the source of muscular energy. *USDA Exp. Station Rec.* *7*:538–50.

Index

Abbreviations: FAO, Food and Agriculture Organization of the United Nations; FPC, fish protein concentrate; H, hydrogen; INCAP, Institute of Nutrition for Central America and Panama; N, nitrogen; PAG, Protein Advisory Group; PCM, protein–calorie malnutrition; RDA, recommended dietary allowance; UN, United Nations Organization; UNICEF, United Nations International Children's Emergency Fund; WHO, World Health Organization of the United Nations

Index

Index

Printed in the United States
By Bookmasters